5th International Workshop on Statistical Physics and Mathematics for Complex Systems (SPMCS 2017)

Journal of Physics: Conference Series
Volume 1113

Wuhan, China
12 - 15 October 2017

Editors:

Wei Li
Sumiyoshi Abe
Alexandre Wang

ISBN: 978-1-5108-8014-6
ISSN: 1742-6588

Printed from e-media with permission by:

Curran Associates, Inc.
57 Morehouse Lane
Red Hook, NY 12571

proceedings.com

Some format issues inherent in the e-media version may also appear in this print version.

Printed by Curran Associates, Inc. (2019)

For permission requests, please contact the Institute of Physics
at the address below.

Institute of Physics
Dirac House, Temple Back
Bristol BS1 6BE UK

Phone: 44 1 17 929 7481
Fax: 44 1 17 920 0979

techtracking@iop.org

Additional copies of this publication are available from:

Curran Associates, Inc.
57 Morehouse Lane
Red Hook, NY 12571 USA
Phone: 845-758-0400
Fax: 845-758-2633
Email: curran@proceedings.com
Web: www.proceedings.com

Table of contents

Volume 1113

5th International Workshop on Statistical Physics and Mathematics for Complex Systems (SPMCS2017) 12–15 October 2017, Wuhan, China

Accepted papers received: 17 October 2018 Published online: 20 December 2018

Prelims

Papers

IOP Conf. Series: Journal of Physics: Conf. Series **1113** (2018) 011001 doi:10.1088/1742-6596/1113/1/011001

Preface

This volume contains 21 contributed papers presented at the *5th International Workshop on Statistical Physics and Mathematics for Complex Systems* (*SPMCS*) held during October 12-15, 2017 in Wuhan, China. Each paper was peer-reviewed by at least one referee chosen from a distinguished international panel.

The previous four workshops of this series were organized in 2008, 2010, 2012, and 2014 in Le Mans of France, Wuhan of China, Kazan of Russia, and Yichang of China respectively.

The SPMCS international workshop series is destined mainly to communicate and exchange research results and information on the fundamental challenges and questions in the vanguard of statistical physics, thermodynamics and mathematics for complex systems. Topics to be covered but not limited to:

• Fundamental aspects in the application of statistical physics and thermodynamics to complex systems and their modeling

• Finite-size and nonextensive systems

• Fluctuation theorems and equalities, quantum thermodynamics

• Variational principle and optimization methods for random dynamics

• Social, economic, biological and technological network structure and dynamic processes

• Equilibrium and non-equilibrium phase transitions

• Temporal network and multi-layer network

• Dynamic processes on complex networks

• Metaphysics and philosophical reflection on the foundation of physics in general

More than 90 participants from 8 countries participated in SPMCS-2017. 47 oral contributions were presented at the workshop.

We would like to take this opportunity to thank the members of the Scientific Program Committee, many of whom acted as reviewers of the papers and responded promptly. We would also like to thank all the guest editors who have done excellent review jobs. Special thanks shall go to the organizing committee, the session chairs, the technicians and the students for the smooth running of the whole workshop. The workshop was also partially supported by Central China Normal University and the Programme of Introducing Talents of Discipline to Universities under grant NO. B08033. Special thanks are due to Ms Juy Zhu who has done excellent editing work with great effort.

Editors: Wei Li, Sumiyoshi Abe, and Alexandre Wang

IOP Conf. Series: Journal of Physics: Conf. Series **1113** (2018) 011002 doi:10.1088/1742-6596/1113/1/011002

Peer review statement

All papers published in this volume of *Journal of Physics: Conference Series* have been peer reviewed through processes administered by the proceedings Editors. Reviews were conducted by expert referees to the professional and scientific standards expected of a proceedings journal published by IOP Publishing.

IOP Conf. Series: Journal of Physics: Conf. Series **1113** (2018) 012001 doi:10.1088/1742-6596/1113/1/012001

Bayesian approach to extreme-value statistics based on conditional maximum-entropy method

Sumiyoshi Abe[1, 2, 3]

[1] Physics Division, College of Information Science and Engineering, Huaqiao University, Xiamen 361021, China

[2] Department of Physical Engineering, Mie University, Mie 514-8507, Japan

[3] Institute of Physics, Kazan Federal University, Kazan 420008, Russia

E-mail: suabe@sf6.so-net.ne.jp (S. Abe)

Abstract. Recently, the conditional maximum-entropy method (abbreviated as C-MaxEnt) has been proposed for selecting priors in Bayesian statistics in a very simple way. Here, it is examined for extreme-value statistics. For the Weibull type as an explicit example, it is shown how C-MaxEnt can give rise to a prior satisfying Jeffreys' rule.

Understanding catastrophic events is a major goal of science of complex systems/phenomena. Examples include megaquakes in seismicity, crashes of financial markets, floods and so forth. A relevant mathematical concept may be extreme-value statistics [1]. Although occurrences of extreme events are rare, systems are strongly characterized by them, actually. It is noted that no big data are available about them, in general. This fact makes the Bayesian approach [2] attractive.

In the Bayesian inference scheme under the circumstance of limited data and/or uncertain knowledge, a key issue is how to select a prior. A lot of investigations have been devoted to this problem in the literature. From the viewpoint of physics, the efforts made by Jeffreys [3] and Jaynes [4] are obviously of particular interest since they base their selection rule upon the invariance principle in "information geometry", which somewhat reminds one of the measure of integration over spacetime in general relativity. In probability theory, however, a distribution is not invariant but accompanied by the Jacobian factor associated with a transformation of variables.

In a recent paper [5], we have formulated a new method of selecting a prior, which is referred to as the conditional maximum-entropy method and is abbreviated as C-MaxEnt. This method has originally been motivated by nonequilibrium statistical mechanics of complex systems governed by hierarchical dynamics with largely-separated time scales [6, 7, 8]. It enables one to express in a succinct way a prior probability merely in terms of a likelihood. It has been found [5] that C-MaxEnt can reproduce the classical results known for the scale and location variables and predicts an intriguing crossover-type (i.e., double power-law) prior for the Poissonian parameter.

In this article, we would like to examine C-MaxEnt for extreme-value statistics. In particular, we analyze Weibull statistics as an explicit example, in detail. We show how C-MaxEnt gives rise to a prior that is consistent with Jeffreys' rule.

Let us start our discussion with an explanation of the procedure of C-MaxEnt. A crucial point of importance is to identify a *conjugate pair of variables* (which should not be confused with conjugate priors). For this purpose, consider a couple of random variables, X and Θ , the realizations of which are denoted by x and θ, respectively. Each of them can be either continuous or discrete, but henceforth both of them are taken to be continuous. The joint probability distribution $P(x, \theta)$ is factorized as follows:

$$P(x, \theta) = p(x|\theta) f(\theta), \tag{1}$$

where $p(x|\theta)$ is the conditional probability distribution, i.e., a likelihood, and $f(\theta)$ is the marginal probability distribution, i.e., a prior. (X, Θ) is referred to as a conjugate pair, if the conditional variance of X behaves as

$$(\Delta X)^2 \propto \frac{1}{\theta^2}. \tag{2}$$

If $p(x|\theta)$ does not have the second moment like the Lévy distribution, then the scaling property or the half width can be used instead of the variance [5]. In the case when there are more variables or odd number of variables, conjugate pairs may not be identified for some. In such a case, it is necessary to find conjugate pairs as many as possible (for details, see Ref. [5]). In view of Eq. (2), the following measure is defined:

$$d\mu = \left(\frac{dx}{k}\right)\left(\frac{d\theta}{l}\right). \tag{3}$$

Here, k and l are coarse-graining factors having the dimensions of X and Θ. For a continuous random variable, such a factor has to be introduced in order for the differential entropy to be definable. For our purpose, however, it does not actually play any role. Therefore, henceforth these factors are simply set equal to unity. Accordingly, the random variables are regarded to be dimensionless. Thus, the differential entropy is defined by

$$S[X, \Theta] = -\int dx d\theta P(x, \theta) \ln P(x, \theta). \tag{4}$$

From the physical viewpoint, Eq. (1) implies existence of the hierarchical structure in the "dynamics" of X and Θ : X is a fast variable, whereas Θ is a slow one. This becomes manifest if Eq. (1) is used for rewriting the joint entropy in Eq. (4) as follows:

$$S[X, \Theta] = S[X|\Theta] + S[\Theta], \tag{5}$$

where $S[X|\Theta]$ and $S[\Theta]$ are the conditional and marginal entropies defined by

$$S[X|\Theta] = \int d\theta f(\theta) S[X|\theta], \qquad S[X|\theta) \equiv -\int dx p(x|\theta) \ln p(x|\theta), \tag{6}$$

$$S[\Theta] = -\int d\theta f(\theta) \ln f(\theta), \tag{7}$$

respectively. It is noted that in Eq. (6) the fast variable is in fact integrated out (i.e., eliminated) first and then $S[X|\theta)$ as a function of θ is averaged afterward. This is in accordance with the way of treating hierarchical dynamics with largely-separated time scales in physics [8]. Actually, Eq. (5) is nothing but Khinchin's second axiom for the Shannon entropy in the discrete case [9].

C-MaxEnt for selecting a noninformative prior is now formulated as follows:

$$\delta_f \left\{ S[X, \Theta] - \lambda \int d\theta f(\theta) \right\} = 0. \tag{8}$$

That is, the joint entropy is maximized only with respect to the prior, i.e., the slow degree of freedom, after elimination of the fast one. λ stands for the Lagrange multiplier associated with the constraint on "normalization" condition on $f(\theta)$. Noninformativeness means that no constraints other than this constraint are imposed. In Bayesian statistics, $f(\theta)$ is often an improper prior, and therefore its integral over the whole range tends to diverge. This is why we do not specify in Eq. (8) the value of the integral of the prior (although in any way such a value turns out to be irrelevant). Using Eqs. (5)-(7) in Eq. (8), we obtain the following main result:

$$f(\theta) \propto \exp\{S[X|\theta)\}. \tag{9}$$

This shows how a noninformative prior can be expressed in terms only of a likelihood.

Now, let us apply C-MaxEnt to the problem of selecting a prior for extreme-value statistics [1]. Let us take N i.i.d. random variables, $X_1, X_2, \ldots, X_N$, and consider the quantity, $(M_N - m_N)/\sigma_N$ with $\sigma_N \in (0, \infty)$ and $m_N \in (-\infty, \infty)$ being constants, where $M_N \equiv \max\{X_1, X_2, \ldots, X_N\}$. The Trinity Theorem states that there exist nontrivial domains of attraction in the limit $N \to \infty$, which lead to extreme-value statistics of the Gumbel, Weibull and Fréchet types. Since general discussions about priors for scale σ_N and location m_N have already been developed in Ref. [5], these are respectively set equal to unity and zero, for the sake of simplicity. Among the three types, here let us treat the one of Weibull as a typical explicit example. The Weibull distribution is given by

$$p(x|\alpha) = \frac{1}{\alpha}(-x)^{1/\alpha - 1} \exp\left[-(-x)^{1/\alpha}\right], \tag{10}$$

where

$$x \in (-\infty, 0], \qquad \alpha \in (0, \infty). \tag{11}$$

Our task is to perform Bayesian inference of the value of the parameter α through its prior by the use of C-MaxEnt, in which $p(x|\alpha)$ in Eq. (10) is regarded as the likelihood. Following the procedure of C-MaxEnt, let us find a conjugate pair of variables. The first and second moments are given by $\langle X \rangle = \int_{-\infty}^{0} dx\, x p(x|\alpha) = -\Gamma(\alpha+1)$ and $\langle X^2 \rangle = \Gamma(2\alpha+1)$, respectively, where $\Gamma(s)$ is the gamma function. So, the conditional variance reads

$$(\Delta X)^2 = \Gamma(2\alpha+1) - [\Gamma(\alpha+1)]^2. \tag{12}$$

This implies that the variable conjugate to X is

$$\Theta = \left\{\Gamma(2A+1) - [\Gamma(A+1)]^2\right\}^{-1/2}, \tag{13}$$

where A is the random variable whose realization is α. $d\theta$ and $d\alpha$ are thus related to each other through the Jacobian factor as follows:

$$d\theta = d\alpha \frac{\psi(2\alpha+1)\Gamma(2\alpha+1) - \psi(\alpha+1)\,[\Gamma(\alpha+1)]^2}{\left\{\Gamma(2\alpha+1) - [\Gamma(\alpha+1)]^2\right\}^{3/2}}, \tag{14}$$

where $\psi(s)$ is the digamma function. On the other hand, $S[X|\alpha)$ defined in the way similar to Eq. (6) is calculated to be

$$S[X|\alpha) = \ln\alpha + \gamma(1-\alpha) + 1, \tag{15}$$

where $\gamma = 0.57721\ldots$ is Euler's constant. Therefore, the prior is found to be

$$f(\alpha) \propto \alpha e^{-\gamma\alpha} \frac{\psi(2\alpha+1)\Gamma(2\alpha+1) - \psi(\alpha+1)\,[\Gamma(\alpha+1)]^2}{\left\{\Gamma(2\alpha+1) - [\Gamma(\alpha+1)]^2\right\}^{3/2}}. \tag{16}$$

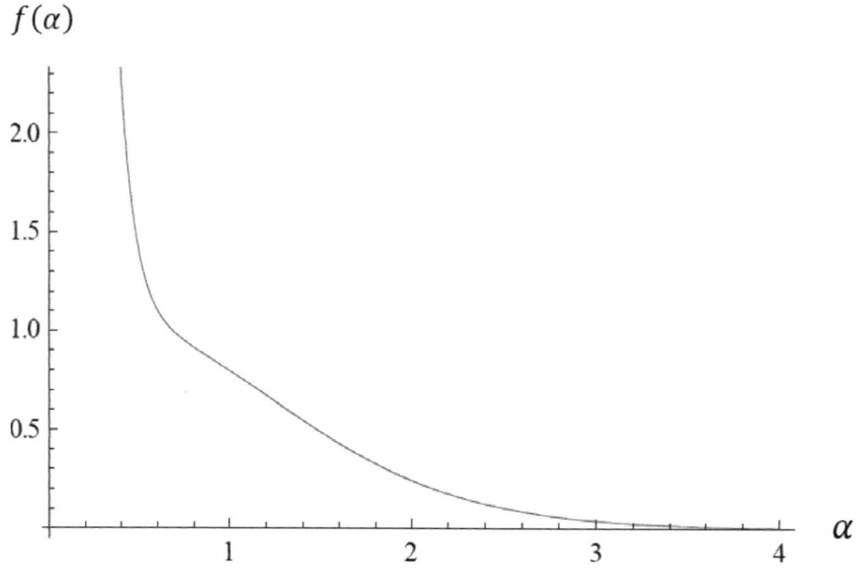

Figure 1. Plot of $f(\alpha)$ in Eq. (16) with respect to α, where the proportionality constant is set equal to unity. All quantities are dimensionless.

The plot of $f(\alpha)$ is presented in figure 1. The asymptotic behaviors of $f(\alpha)$ are analytically evaluated as follows:

$$f(\alpha) \sim \frac{1}{\alpha} \quad (\alpha \to 0+), \qquad \exp(-\alpha \ln \alpha) \quad (\alpha \to \infty). \tag{17}$$

Note that the linear divergence in the limit $\alpha \to 0+$ is plausible in view of Jeffreys' rule as discussed by Jaynes in Ref. [4].

In conclusion, we have applied C-MaxEnt to the problem of selecting a prior in Bayesian approach to extreme-value statistics. Taking the Weibull type as an explicit example, we have shown how C-MaxEnt can yield a prior that is reasonable in view of Jeffreys' rule.

Acknowledgments

This work has been supported in part by grants from National Natural Science Foundation of China (No. 11775084) and Grant-in-Aid for Scientific Research from the Japan Society for the Promotion of Science (No. 26400391 and No. 16K05484), and by the Program of Competitive Growth of Kazan Federal University from the Ministry of Education and Science of the Russian Federation.

References

[1] Coles S 2001 *An Introduction to Statistical Modeling of Extreme Values* (Berlin: Springer-Verlag)
[2] Berger J O 1993 *Statistical Decision Theory and Bayesian Analysis* 2nd edition (New York: Springer-Verlag)
[3] Jeffreys H 1946 *Proc. R. Soc. London Ser.* A **186** 453
[4] Jaynes E T 1968 *IEEE Trans. Syst. Sci. Cybern.* **SSC-4** 227
[5] Abe S 2014 *EPL* **108** 40008
[6] Beck C and Cohen E G D 2003 *Physica* A **322** 267
[7] Abe S, Beck C and Cohen E G D 2007 *Phys. Rev.* E **76** 031102
[8] Abe S 2014 *Found. Phys.* **44** 175
[9] Khinchin A I 1957 *Mathematical Foundations of Information Theory* (New York: Dover)

IOP Conf. Series: Journal of Physics: Conf. Series **1113** (2018) 012002 doi:10.1088/1742-6596/1113/1/012002

Universality and scaling in human and social systems

Chin-Kun Hu[1,2]

[1]Institute of Physics of Academia Sinica, Taipei 11529, Taiwan
[2]Department of Physics, National Dong Hwa University, Hualien 97401, Taiwan

E-mail: `huck@phys.sinica.edu.tw`

Abstract. The objective of statistical physics is to understand macroscopic behavior of a many-body system from the interactions of the constituents of that system. When many-body systems reach critical states, simple universal and scaling behaviors appear. In this talk, I first introduce the concepts of universality and scaling in critical physical systems, I then briefly review some examples of universal and scaling behaviors in human and social systems, e.g. universal crossover behavior of stock returns, universality and scaling in the statistical data of literary works, universal trend in the evolution of states or countries etc. Finally, I mention some interesting problems for further studies.

1. Introduction

"Universality" means applicable in different content; "scaling" means physical quantities from different system sizes as functions of the parameter, e.g. temperature, of the systems can collapse approximately on a single curve by a simple change of scale for each quantity [1]. A good example of scaling is Russian doll, which contains a set of self-similar dolls of different length scales. One of the objectives of scientific research is to use a small number of concepts, or functions, or laws to describe a large numbers of natural or laboratory phenomena. Thus it is not surprising that the concepts of universality and scaling have been used by researchers in different branches of sciences to describe behavior of the studied systems [1].

Universality is the key idea in the search for fundamental interactions of matter. Issac Newton (1642-1727 A.D.) discovered that the falling of apples to the earth, the motion of the moon around the earth, and the motion of the earth around the sun are governed by the same physical laws: universal law for gravitational forces of matter and Newton's laws of motion. James Clerk Maxwell (1831-1879 A.D.) proposed Maxwell's equations to describe electric and magnetic interactions. In recent decades, S. Weiberg and A. Salam proposed a unified gauge field theory to describe the weak and electromagnetic interactions. The major objective of theoretical elementary particle physics is to formulate a unified theory which can describe electromagnetic, weak, strong, and gravitational interactions [1].

Besides basic interactions, the idea of universality is also useful for the understanding of many-body interacting systems, especially the critical systems [2, 3]. The quantitative and systematic researches on critical physical systems (e.g. CO_2 and H_2O) started in the later half of the 19th. century in the study of liquid-gas critical systems, and made tremendous progress in the 20th. century via the ideas and methods from statistical physics [2, 3].

In 1869, Thomas Andrews (9 December 1813 - 26 November 1885) presented a detailed report on the critical phenomena of CO_2 [4]. The data in [4] were used by S. G. Brush to plot isothermal

IOP Conf. Series: Journal of Physics: Conf. Series **1113** (2018) 012002 doi:10.1088/1742-6596/1113/1/012002

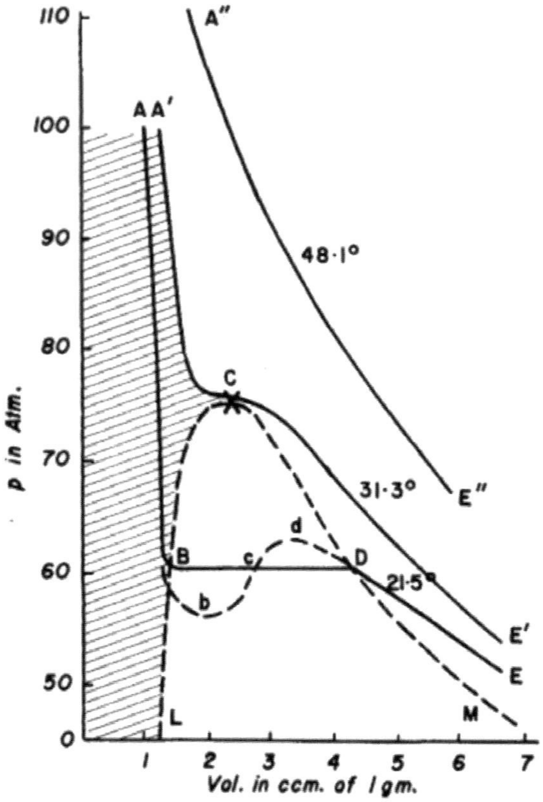

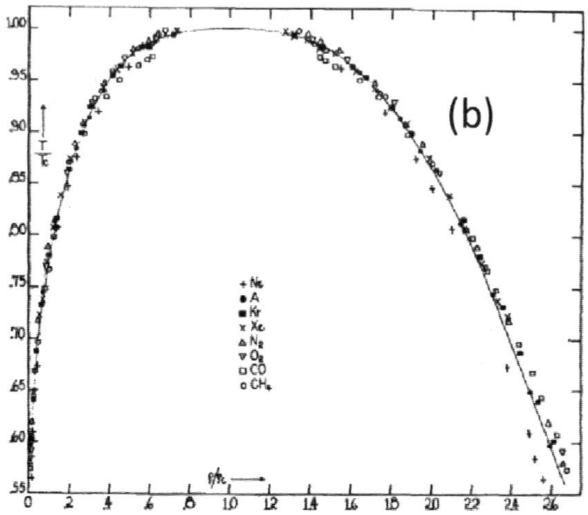

(a) Taken from S. G. Brush, The Kind of Motion We Call Heat, 1 (North-Holland, Amsterdam, 1976).
(a) Taken from E. A. Guggenheim, The Journal of Chemical Physics 13, 253 (1945).

Figure 1. (a) Isothermal curves for CO_2 in the pressure (P) vs. volume (V) plane plotted by S. G. Brush [5] who used the data reported in [4] by Thomas Andrews in 1869. (b) Collapse of data for 8 gas-liquid systems on the universal curve with a critical exponent 1/3. Taken from [10] by E. A. Guggenheim.

curves of CO_2 on the pressure P vs. colume V plane [5], which is reproduced in Fig. 1(a).

In 1834, Emile Clapeyron (1799-1864) proposed the ideal gas law $PV = RT$ for a mole ($\approx 6.02 \times 10^{23}$ molecules or atoms) of ideal gas, where R is the ideal gas constant, and T is the absolute temperature. In 1873 Johannes Diderik van der Waals (23 November 1837 - 8 March 1923) took into account finite sizes and interactions of gas molecules to modify the ideal gas law and proposed that a mole of liquid-gas system can be described by the equation [2, 3, 6, 7]

$$\left(P + \frac{a}{V^2}\right)(V - b) = RT, \tag{1}$$

where a and b are constants. One can use Eq. (1) to plot P as a function of V for a fixed T in the P-V plane. For a constant $T < T_c$ and a constant P, Eq. (1) could have three solutions for V; in this case Maxwell construction (an equal area rule for the parts below and above horizontal line) can be used to obtain a curve in the P-V plane, which has a horizontal portion corresponding to liquid-gas coexistence region [2]. When T is increased, the size of this region decreases. When T reaches the critical temperature T_c, the size of this region becomes 0. From this condition, one can derive that [2, 3] $V_c = 3b$, $P_c = a/27b^2$, $T_c = 8a/27bR$, from which one can determine a, b, and ideal gas constant R from measured values of V_c, P_c, and T_c. Two curves which connect boundaries of the horizontal portions for $T \leq T_c$ form phase boundaries for liquid phase, liquid-gas coexistence phase, and gas phase. For $T \gg T_c$, Equation (1) can be approximated by the ideal gas law $PV = RT$.

In 1881, van der Waals defined reduced volume ϕ, reduced pressure π, and reduced temperature θ as $\phi = V/V_c$, $\pi = P/P_c$, $\theta = T/T_c$, then he rewrote Eq. (1) as

$$\left(\pi + \frac{3}{\phi^2}\right)(3\phi - 1) = 8\theta, \tag{2}$$

which is called reduced van der Waals' equation and means that in terms of reduced variables, data for different substance can collapse on the same curve. In 1910, van der Vaals received the Nobel Prize in physics "for his work on the equation of state for gases and liquids".

In 1924-1931, Lennard-Jones [8, 9] proposed that the interaction potential between a pair of atoms or molecules at sites i and j with the distance r_{ij} is

$$v_{ij}(r_{ij}) = 4\epsilon\left(\left(\frac{\sigma}{r_{ij}}\right)^{12} - \left(\frac{\sigma}{r_{ij}}\right)^{6}\right), \tag{3}$$

where σ and ϵ are used as units for length and energy strength, respectively. Thereafter, the potential of Eq. (3) is called Lennard-Jones potential.

In 1945, Guggenheim [10] reported that in the T/T_c versus ρ/ρ_c plane (ρ and ρ_c are density and critical density of the substance, respectively) the coexistence curves of N_e, A_r, K_r, X_e, N_2, O_2, CO, and CH_4 are very consistent with each other, which is reproduced in Fig. 1(b). Thus van der Waals's general idea for the universality of the coexistence curves was confirmed. However, near the critical point, Guggenheim found that

$$\frac{\rho_l - \rho_g}{\rho_c} = \frac{7}{2}|1 - \frac{T}{T_c}|^{\beta}, \tag{4}$$

where the critical exponent β is 1/3 [10], on the other hand van der Waals's equation gives $\beta = 1/2$ [2]. The discovery of such deviation marks modern era of critical phenomena [2].

In1952, Yang and Lee proposed that the critical behavior of the gas-liquid system can be represented by a lattice-gas model [11], which is equivalent to the Ising model [12], in which each atom or molecule on a lattice site is assigned a variable which can be either +1 or -1.

In 1960s, many experimental and theoretical results indicate the universality and scaling in critical physical systems [2, 3]. It has been proposed that critical systems can be classified into different universality classes so that the systems in the same class has the same set of critical exponents. It has been found that critical exponents satisfy certain equalities, call scaling laws or relations [2, 3]. The phenomenal theory had been proposed to understand the scaling relations [2, 3]. In 1971-1972, Ken G. Wilson developed renormalization group (RG) theory [13, 14] for critical phenomena, which provides a framework for understanding universality and scaling in critical systems, and can be used to calculate critical quantities of critical systems. For this contribution, Ken G. Wilson won the Nobel Prize in physics in 1982 [15].

In 1995-1996, Blöte and collaborators [16, 17] used Monte Carlo simulations to find that the critical exponent β of the spontaneous magnetization and ν of the correlation length [2] of a three-dimensional (3D) Ising model are 0.3269(6) [17] and 0.6301(8) [16], respectively. In 2009, Sengers and Shanks [18] reviewed the experimental data for liquid-gas critical systems and reported that $\beta = 0.3245$ and $\nu = 0.629 \pm 0.003$, respectively. In 2012, Watanabe, Ito and Hu [19] used molecular dynamics simulations to find that β and ν of a 3D Lennard-Jones (L-J) model system [8, 9] are 0.3285(7) and 0.63(4), respectively. The values of β and ν obtained for simple model systems reported in Refs. [16, 17, 19] are highly consistent with experimental data

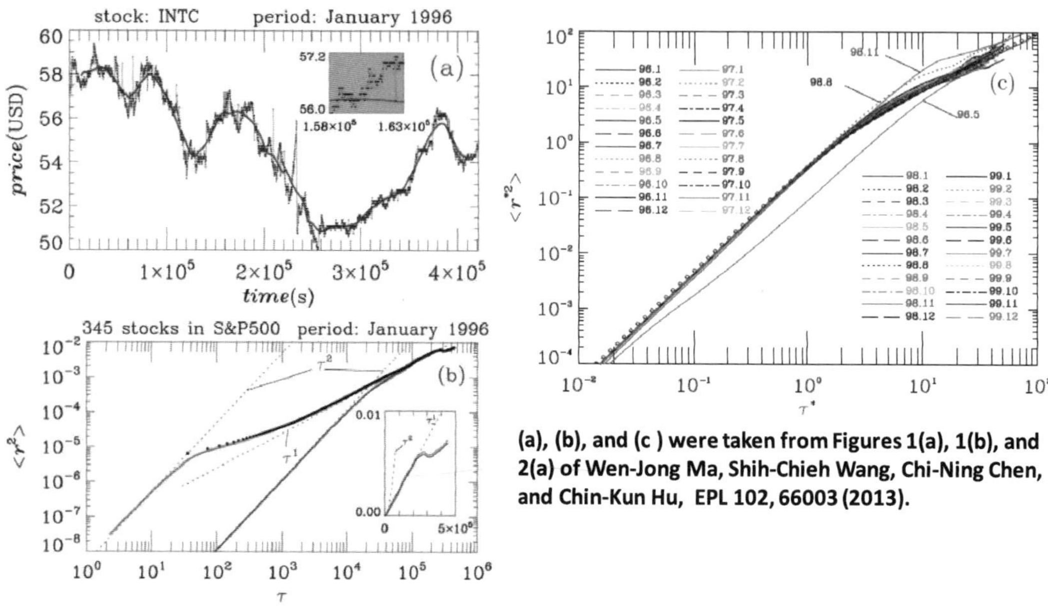

(a), (b), and (c) were taken from Figures 1(a), 1(b), and 2(a) of Wen-Jong Ma, Shih-Chieh Wang, Chi-Ning Chen, and Chin-Kun Hu, EPL 102, 66003 (2013).

Figure 2. (Color online)(a) Stock prices of Intel Corporation (INTC), over 18 trading days in January 1996. The market data, collected in 36-seconds intervals, are marked by dots, which are connected by red lines and their one-day moving averages in steps of 36-seconds are plotted by the blue line. The colored inset shows the enlarged image of the region marked by the same (green) color. (b) Log-log plot for the mean-square-log-returns (MSLR) versus time interval τ, for a collection of 345 stocks as in [55], including INTC, picked from S&P500 over January 1996, calculated either based on original data (marked by dot points), or based on the moving-average (blue curve) shown in (a). The red line shows the result over short time scales, based on the continuous (red) curve of (a); the blue curve shows the results obtained from the blue curve in (a). In the inset, we show the plots the two former curves in linear scales. (c) Scaled mean square log-return $< r^{*2} >$ versus scaled time internal τ^*, for each month in 1996-1999, for the 345 stocks in S&P500 in US that have been analyzed in (b). They are compared with the master curve (open circles) defined by Eq. (7). The time sequences for the collections of data are in intervals of 36 seconds. The fitting to asymptotic form θt_1^α and Dt are carried out over the ranges $36 < \tau < 7200$ and $54000 < \tau < 108000$. Those months with their scaled MSLRs significantly deviating from the master curve are marked specifically.

reported in [10, 18]. Using Monte Carlo methods [20, 21, 22] and analytic methods [23, 24] it has been found that many critical systems can have very nice universal and scaling behaviors [25, 26, 27, 28, 29, 30, 31, 32, 33, 34].

It is of interest to know whether the concepts and methods from critical physical systems can be useful for understanding complicated biological and human-social systems. In previous papers [35, 36, 37, 38, 39, 40, 41, 42, 43], we had pointed out that the ideas and methods of statistical physics can be useful for understanding some interesting biological problems. In this paper, I would like to point out that such methods and concepts can also be useful for understanding some interesting human and social systems.

2. Universality and scaling in stock markets

In 1828, R. Brown published a paper about "A brief account of microscopic observations on the particles contained in the pollen of plants and on the general existence of active molecules" [44].

In 1900, Louis Bachelier (1870-1946) proposed a random walk model for financial market [45]. In 1905-1908, Albert Einstein (1879-1955) published a series of papers about random walk model for Brownian motion [44] in Germany whose English translation can be found in [46].

In 1908, Paul Langevin (1872-1946) proposed a stochastic equation to describe the displacement of Brownian particles [47] in French whose English translation was published in [48]. Later studies indicate that the financial market does not follow the random walk model [45]. Some results from [49] by Ma, Wang, Chen and Hu are mentioned below.

Mantegna and Stanley [50] showed that the scaling of the probability distribution of an economic index (S & P 500) can be described by a non-gaussian process for data in 1984-89. Scaling behavior is observed for time intervals from 1,000 min to 1 min. Saakian *et al.* obtained exact non-Gaussian distribution of stock returns from the multifractal random walk model [51].

In 1999, results from two groups indicate that the evolution of stocks can not be described by the random walk model [45]. Laloux, *et al.* [52] calculated correlation matrix for 406 stocks in S & P 500 during 1991-1996 with time interval of one day and Plerou, *et al.* [53, 54] calculated correlation matrix for 1000 stocks in USA during 1994-1995 with time interval of 30 minutes. Both groups found that in the eigenvalue distribution of the correlation matrix, there are some discrete distributed larger eigenvalues above the continuous component predicted by the random walk model for stocks. In the eigenvector corresponding to the largest eigenvalue of the correlation matrix (λ_M), all stocks in the market move (deviate from the average value) in the same direction. The mode corresponding to λ_M is called the market mode.

After the work reported in [52] and [53, 54], one interesting question was how to modify the random walk model so that the revised model can well describe the eigenvalues from stock markets. For this purpose, Ma, Hu and Amritkar [55] proposed a model of coupled random walks for stock-stock correlations (see also [56]); the walks are coupled via a mechanism that the displacement (price change) of each walk (stock) is activated by the price gradients over some underlying network. They assumed that the network has two underlying structures: one for the correlations among the stocks of the whole market and another for those within individual groups; each with a coupling parameter controlling the degree of correlation. The model provides the interpretation of the features displayed in the distribution of the eigenvalues for the correlation matrix of real market on the level of time sequences. They verified that such a model indeed reproduces the major eigenvalue spectra of the US stocks.

The mean squares of log-return for stocks are known to have an asymptotic τ^α dependence on the time interval τ for large τ, with the exponent α close to unity. Such a property is similar to the diffusion behavior of particles in their mean square displacement [49]. It is of interest to know whether such analogy can be extended to short τ region. To answer this question, we study the average square log return with different time interval τ.

Figure 2(a) shows the stock price $P(t)$ of the INTEL Corporation as a function of time t, for 18 trading days in January 1996; each day has trading time 6.5 hours=23400 seconds. For the convenience of analysis, the data are collected in the time interval of 36 seconds. For the price changes of a collection of stocks, we consider the analogy between the particle displacement and the log-return $r(t,\tau) = \log(P(t+\tau)) - \log(P(t)) \approx (P(t+\tau) - P(t))/P(t)$ for the price $P(t)$ over an interval τ starting at time t. The log-return carries the changes relative to the prices so that it is a quantity that effectively renders all stocks on an equal foot, despite of the inherited heterogeneities among the companies and their stocks prices. The dot points in Fig. 2(b) shows the mean square log-return (MSLR) $< r^2 >$ of a collection of 345 stocks (the same as those studied in [55]) of the S&P 500, versus τ during the month January of 1996. The average $< \cdot >$ is taken over all the events for any time interval $(t, t+\tau)$ from the 36-second-step data, for any of the 345 stocks. The result does share the major features with the mean square displacement (MSD) of particles, that makes possible an effective comprehension of these empirical data.

To reveal the analogy in time evolution between the stock prices and the particle trajectories, we analyze the paths of the "motion" of the stocks in the one dimensional price space. We adopt two different ways to analyze the data, either by refining the intra-day microscopic temporal features or by eliminating those heterogeneity via a coarse-graining procedures. For the latter, we collect the high frequency one-day moving averages (HF1MV) $\bar{P}(t)$ of the prices for individual stocks, by taking simple averages over a shifting window of one-trading-day wide (23400 seconds, or 650 intervals). To retain the high frequency feature of the data, the window shifts in steps of 36 seconds. The blue line in Fig. 2(b) shows the 36-second MSLR, $< r^2 >$, of January 1996, calculated from the returns $r(t, \tau) = \log(\bar{P}(t + \tau)) - \log(\bar{P}(t))$ for the collection of 345 stocks.

A simple model to cover both time regimes, for the motion of a tracer particle of mass m in velocity $v(t) = \frac{dR}{dt}$ in $n_d = 1$ dimension, is described by the Langevin equation [47, 48] of motion with white noise $\xi(t)$

$$\frac{dv}{dt} = -\frac{1}{\tau_0} v(t) + \xi(t). \tag{5}$$

The balance between the friction $\frac{m}{\tau_0} v(t)$ and the random force $m\xi(t)$ imposes a condition on the amplitude of $\xi(t)$: $< \xi(t)\xi(t') >= \frac{2m^2 v(0)^2}{\tau_0} \delta(t - t')$. The MSD is given by [57]

$$\text{MSD} \equiv < (R(t + \tau) - R(t))^2 >= D[\tau - \tau_0(1 - e^{-\tau/\tau_0})] \tag{6}$$

where $D = 2\tau_0 v(0)^2$. For the smaller and the larger τ, we have asymptotic MSD $\approx \frac{D}{2\tau_0}\tau^2$, for $\tau \approx 0$, and MSD $\approx D\tau$, for $\tau >> 1$, respectively. Note that the ratio μ between the pre-factor D of the latter asymptotic expression to that $\frac{D}{2\tau_0}$ of the former one measures the "mobility" of the tracer particle. The result $\mu = 2\tau_0$ is Einstein's relation.

The averaging procedure adopted in HF1MV effectively eliminates the stretched crossover between the two asymptotic regimes in the MSLR (dot points and blue line in Fig. 2(b)), so that the data can be described by the simple Langevin equation (Eq. (6)). To facilitate such a conjecture, we use Eq. (6) as a guide to write the master equation

$$< r^{*2} >= \tau^* - 1 + e^{-\tau^*} \tag{7}$$

for fitting the scaled MSLR $< r^{*2} >=< r^2 > /(D\tau_0)$ versus $\tau^* = \tau/\tau_0$ for the empirical data. The data are well-fitted by $< r^2 >\approx \theta\tau^{\alpha_1}$ over the smaller τ regime to obtain the pre-factor $\theta = \frac{D}{2\tau_0}$. We found that the data are well fitted to give $\alpha_1 = 2.0$ with an error less than 1%. For the larger τ regime, the fitting to the asymptotic form $< r^2 >\approx D\tau^{\alpha_2}$ leads to much more scattered values in α_2. To cure this drawback, we impose $\alpha_2 = 1$ and find the best fit for D.

In Figs. 2(C), we show the scaled curves of $< r^{*2} >$ versus τ^*, for the 48 months during the years 1996-1999, for 345 stocks from S&P 500 analyzed in Fig. 2(b). The trading hours for a trading day in 1996-1999 are six and half hours (9:30AM-4:00PM) for the US market. The time sequences for the collections of US market are in intervals of 36 seconds. Treating each market as temporarily stationary over each month, an average $< \cdot >$ is taken both time-wise and stock-wise. In spite of a slight degree of bulge at the crossover, we find that the scaled data are in general in a reasonable agreement with Eq. (7) (Fig. 2(C),), that the fitted parameters θ and D provide useful information about the condition of the market.. In Fig. 2(C), there are a few curves deviating significantly from the master curve. These deviations signal temporarily non stationarity of the markets.

Equation (7) for the scaled empirical data of Fig. 2(c) is very similar to Eq. (4) for the scaled empirical data of Fig. 1(c).

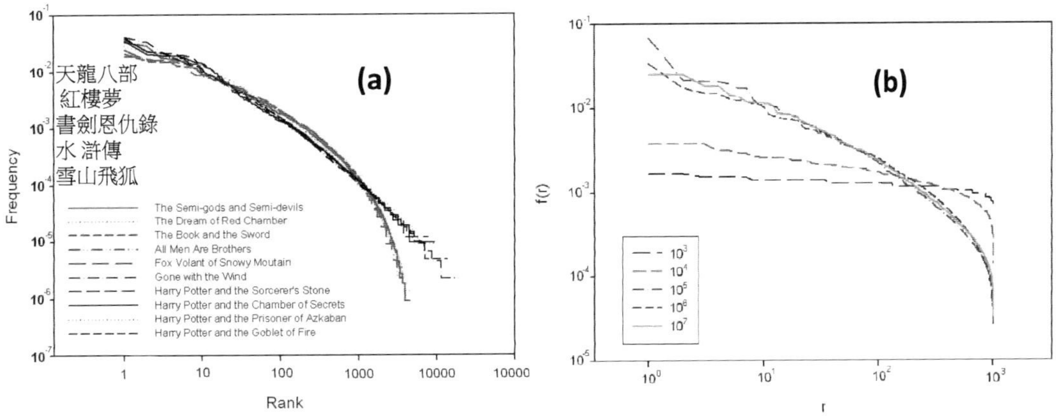

Figure 3. (Color online) (a) Zipf's distributions for five Chinese novels (The Semi-gods and Semi-devils, The Dream of Red Chamber, The Book and the Sword, All Men Are Brothers, and Fox Volant of Snowy Moutain) and five English novels (Gone with the Wind, Harry Potter and the Sorcerer's Stone, Harry Potter and the Chamber of Secrets, Harry Potter and the Prisoner of Azkaban, and Harry Potter and the Goblet of Fire). The Chinese titles of Chinese novels are listed in the same order above the English titles. Please note that data of the former and data of the latter fall on separate curves. (b) An example of computer simulations to generate the Zipf's distribution with time steps 10^3, 10^4, 10^5, 10^6 and 10^7. As time steps increase, the distribution approaches a steady distribution. (a) and (b) were taken from Fig. 8 and Fig. 9 of [1], respectively.

.

3. Universal patterns in human writing

In this section, I briefly introduce the results in Section 5 of [1] by Hu and Kuo.

In 1949, George Zipf published an interesting result about statistical data in human writing [58]. He found that in a given text corpus there is an approximate mathematical relation between the frequency of the occurrence of each word and its rank in the list of all the words in the text ordered by decreasing frequency [58]. The distribution is called the Zipf's distribution.

Zipf's distributions for five Chinese novels and five English novels had been calculated and shown in Fig. 3(a). The Chinese novels were written in Ming Dynasty, Ching Dynasty, and recent decades. It is of interest to know that Zipf's distributions for 5 Chinese novels fall approximately on the same curve and Zipf's distributions for 5 English novels fall approximately on another curve, which is quite different from that of Chinese novels. Thus there are two classes of Zipf's distributions: one for Chinese novels and another one for English novels.

It has been found that critical physical systems can be classified into different universality classes so that the systems in the same class have the same set of critical exponents [2, 3], universal finite-size scaling functions [25, 27, 28, 29, 31, 32, 34], and amplitude ratios [33]. The criteria by which the critical systems can be classified into different universality classes is

IOP Conf. Series: Journal of Physics: Conf. Series **1113** (2018) 012002 doi:10.1088/1742-6596/1113/1/012002

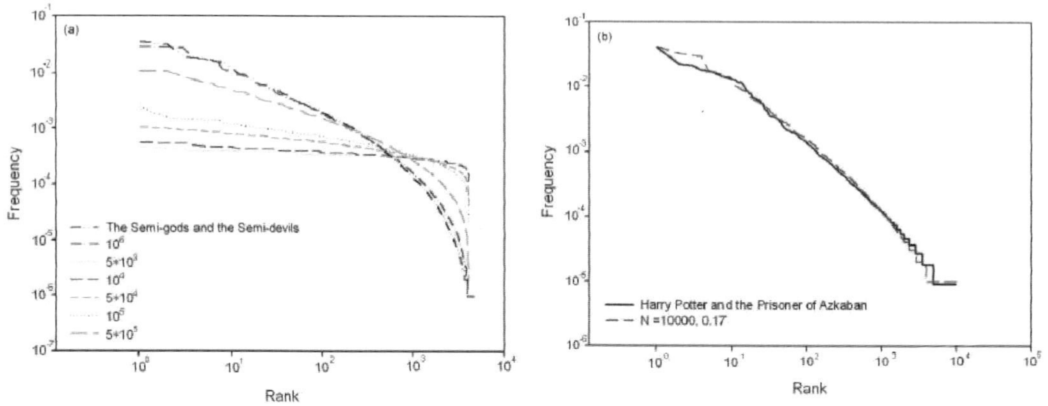

Figure 4. (Color online) Numerical simulations of Zipf's distributions: (a) number of words N =4000, exchange rate $g = 0.085$ and simulation steps: 5×10^3, 10^4, 5×10^4, 10^5, and 10^6. (b) number of words $N = 10000$, exchange rate $g = 0.17$ and simulation steps 10^7. (a) and (b) were taken from Fig. 10(a) and Fig. 10(b) of [1], respectively.

a problem of much academic interest. Figure 3 suggests Chinese literary works and English literary works could be considered as two separate "universality classes". It is of interest to know what are differences between these two classes. In the following, a model proposed by Hu and Kuo [1] to generate such curves is introduced.

A writer uses N different words w_i, $1 \le i \le N$, in his or her writing and the probability that he or she uses w_i in the writing is x_i, which satisfy the normalization relation $\sum_{i=1}^{N} x_i = 1$. We can arrange x_i in a sequence from larger to smaller values and define an integer function $I(x_i)$ of x_i, which is the order number of x_i in the sequence. We assume that all x_i are different and we can define an inverse function V of I such that $y_I \equiv x_i = V(I(x_i))$.

We assume that a mature writer would have a steady x_1, x_2, ..., x_N and the corresponding y_1, y_2, ..., y_N, which was developed following process with an exchange parameter g and a range parameter r.

(i) In the beginning, every x_i takes a random number near $1/N$ such that $\sum_{i=1}^{N} x_i = 1$. From $(x_1,\ x_2,\ \ldots,\ x_N)$, one has a corresponding $(y_1,\ y_2,\ \ldots,\ y_N)$ so that $y_i > y_{i+1}$ for 1 between 1 and $N-1$.
(ii) Randomly choose an integer j between 1 and N, and subtract gy_j from y_j: $y_j \to y_j - gy_j \ge 0$.
(iii) Randomly choose an integer k between $j-r$ and $j+r$ and add gy_j to y_k : $y_k \to y_k + gy_j$.
(iv) Update the order of y_i such that $y_i > y_{i+1}$ for j between 1 and $N-1$.
(v) Go to step 2 and iterate the process.

The simulation results for y_I as a function of I in log-log scale for $r = 3$, $N = 1000$, $g = 0.1$, and iteration number IN $= 10^3$, 10^4, 10^5, 10^6 and 10^7 are presented in Fig. 3(b) which shows

(a) 羅貫中 三國演義 第001回 Luo Guanzhong: Romance of the Three Kingdoms Chapter 001
話說天下(Tien Xia, The world)大勢，分久必合，合久必分。周末七國分爭，併入於秦。及秦滅之後， 楚、漢分爭，又併入於漢。 漢朝自高祖 (Emperor Gaozu of Han) 斬白蛇而起義，一統天下。後來光武(Emperor Guangwu of Han) 中興，傳至獻帝(Emperor Xian of Han)， 遂分為三國(Three Kingdoms)。...

(b) 羅貫中 三國演義 第120回 Luo Guanzhong: Romance of the Three Kingdoms Chapter 120
..... (280AD)孫皓 (Sun Hao) 聞晉兵入城，欲自刎。中書令胡沖、光祿勳薛瑩奏曰：「陛下何不效安樂公劉禪乎？」皓從之，亦輿櫬自縛，率諸文武，詣王濬軍前歸降。濬釋其縛，焚其櫬，以王禮待之。....
自此三國歸於晉帝司馬炎，為一統之基矣。此所謂「天下大勢，合久必分，分久必合」者也。

(C) 班固 漢書 95卷 Ban Gu Book of Han Chapter 095
(漢武帝劉徹 Emperor Wu of Han Liu Che) 元鼎五年(112BC)秋，衛尉路博德為伏波將軍，出桂陽，下湟水；主爵都尉楊僕為樓船將軍，出豫章， 下橫浦；故歸義粵侯二人為戈船、下瀨將軍，出零陵，或下離水，或抵蒼梧；使馳義侯因巴蜀罪人， 發夜郎兵， 下牂柯江：咸會番禺。
(元鼎)六年(111BC)冬，樓船將軍將精卒先陷尋陿，破石門，得粵船粟，因推而前，挫粵鋒，以粵數萬人待伏波將軍。伏波將軍將罪人，.....。粵桂林監居翁諭告甌駱四十餘萬口降，為湘城侯。戈船、下瀨將軍兵及馳義侯所發夜郎兵未下，南粵 (Nanyue) 已平。遂以其地為儋耳、珠崖、南海、蒼梧、鬱林、合浦、交阯 (Jiaozhi) 、 九真 (Jiuzhen)、 日南 (Rinan)九郡(9 commanderies)。伏波將軍益封。樓船將軍以推鋒陷堅為將梁侯。

(d) 班固 漢書 95卷 Ban Gu Book of Han Chapter 095
(漢武帝劉徹)元封三年(108BC)夏，尼谿相參乃使人殺朝鮮王右渠來降。王險城未下，故右渠之大臣成已又反，復攻吏。左將軍使右渠子長、降相路人子最，告諭其民，誅成已。故遂定朝鮮為真番(Zhenfan)、臨屯(Lintun)、樂浪(Lelang Commandery)、玄菟 (Xuantu)四郡 (4 Commanderies) 。

Figure 5. (Color online) (a) Some statements from Luo Guanzhong: Romance of the Three Kingdoms Chapter 001 . (b) Some statements from Luo Guanzhong: Romance of the Three Kingdoms Chapter 120. (c) Ban Gu Book of Han Chapter 095 about Vietnam. (d) Ban Gu Book of Han Chapter 095 about the Korea.

that as IN increases, the Zipf's distributions approach a steady curve similar to the curve for Chinese literary works in Fig. 3(a).

To understand the differences between curves for Chinese and English novels in Fig. 3(a), Hu and Kuo [1] tried to simulate curves in Fig. 3(a) and found that with r=3, N =4000 and g= 0.085 the simulated curves can approach the curves for Chinese novels as shown in Fig. 4(a) and with r=3, N =10000 and g = 0.17, the simulated curves can approach the curve for English novels as shown in Fig. 4(b). Thus the English novels have larger number of different words N and exchange rate g than Chinese novels. This is not surprising because the former language is an alphabet spelling system and it is easier to generate new words to enrich the vocabulary or to use one word to replace another one during development of writing habit.

Zipf's distribution has been used to solve the authorship dispute problem [59, 60]. Shiji prepared by Shima Qian in Western Han Dynasty (202 BC-9 AD) was the most famous historical book in historical China [60] [1] The current version of Shiji contains 130 chapters, including 10 chapters which contains some parts starting with the sentence "Mr. Chu said ..."; Chang Yen in the Three Kingdoms Period (220-280 AD) considered that such parts were written by Chu Shaosun, which was widely accepted. We have analyzed the Zipf distributions in Shiji and books written by Liu Xiang (77 BC-6 BC) in Western Han Dynasty. Our results and other evidences indicate that the parts of Shiji starting with the sentence "Mr. Chu said ..." might be prepared by Liu Xiang [60]. One of the other evidences is Fig. 6 of [60]. The texts in Figs. 6(a) and 6(c) of [60] suggest that Chu Shaosun might not be Mr. Chu in Shiji. After the publication of [60], on 30 May 2017 (Duanwu Festival or Dragon Boat Festival), I learned that Mr. Wang Guowei

[1] In Eq. (2.6) of [60], the sum is over all words w_i in the union of top R_{cut} words in the i-th and j-th texts.

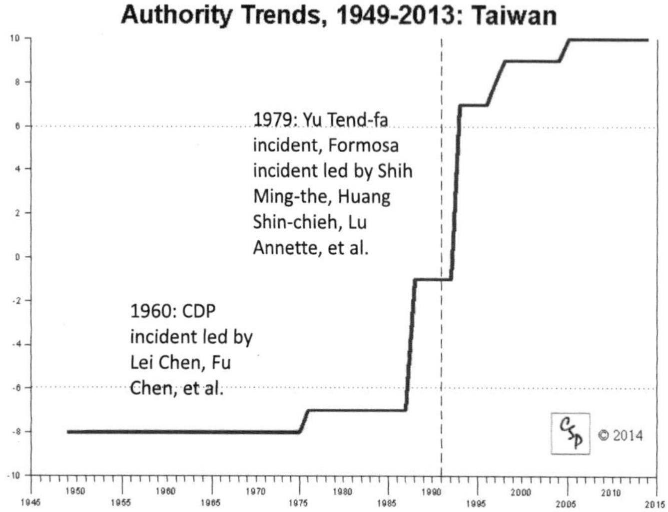

Taken from http://www.systemicpeace.org/polity/polity4x.htm

Key terms & persons related to the left figure:
National Assembly 國民大會
Legislative Yuan 立法院
President 總統
KMT 中國國民黨
Chiang Kai-shek 蔣中正
Chiang Ching-kuo 蔣經國
Lee Teng-hui 李登輝
China Democratic Party中國民主黨
Lei Chen 雷震
Fu Chen 傅正
Yu Teng-fa 余登發
Hsu Hsin-liang 許信良
Shih Ming-the 施明德
Huang Shin-chieh 黃信介
Lu Annette 呂秀蓮
Formosa Incident 美麗島事件
Chen Shui-bian 陳水扁
DPP 民主進步黨

Figure 6. Chiang Kai-shek was elected as President of the Republic of China on 24 April 1948, 21 March 1954, 21 March 1960, 21 March 1966, and 21 March 1972. There was a movement against Chiang Kai-shek's re-election in 1960 led by Lei Chen. After Chiang's re-election in 1960, there was a movement to form a new party: China Democratic Party (CDP) Led by Lei Chen. This movement was suppressed by KMT successfully by putting Lei Chen in jail. In 1977-1990, there was another democratic movement. After many leaders of this movement were put into jails in early 1979 (Yu Deng-fa) and December 1979 (Formosa incident), the movement was continued by family members of the leaders, lawyers, and students, and Taiwan became fully democratic. The event in 1960 is similar to a stable point between A and B in Fig. 1(a). The events in 1979 are similar to point "b" (a meta-stable state) in Fig. 1(a).

(1877-1927) expressed the similar idea in an article "*Han Wei po shi tee ming kao*" which stated that " Chu Shaosun and Mr. Chu might not be the same person."

4. Do countries or states have universal evolution behavior or pattern ?

Countries or states are composed of a large number of people. Do countries or states have universal behavior or pattern in their long time evolution?

Romance of the Three Kingdoms is a historical novel about the Three Kingdoms period in China from Yellow Turban Rebellion in 184 AD in the Eastern Han Dynasty (23-220 AD) to 280 AD when the Emperor of *Eastern Wu* surrendered to the Jin Dynasty (see below). The novel was first published in the Ming Dynasty (1368-1644AD) and it was widely believed that the author of the novel was Luo Guanzhong who was a writer from the end of the Yuan Dynasty (1271-1368 AD) to the beginning of the Ming Dynasty.

The first chapter of the novel contains a statement about the universal pattern for the long time evolution of states or countries: "The major trend of the world (*Tien Xia* meaning under the heaven) is unification after long period of division, and division after long period of unification". To support this statement, it was mentioned that after long warring state period (476-221 BC),

Qin established a unified "Qin Dynasty" (in 221 BC). After the collapse of Qin Dynasty (in 206 BC), Chu and Han competed with each other and Emperor Gaozu of Han (about 256-195 BC) unified the world (in 202 BC) to establish Han Dynasty.... After long period of unification, Han Dynasty collapsed during Emperor Xian of Han (181-234 AD, as emperor in 189-220 AD) and enter Three Kingdoms period, see Fig. 5(a).

Three Kingdoms included *Cao Wei* (220-266 AD) in the Yellow River region, *Shu-Han* (221-263 AD) in Sichuan, and *Eastern Wu* (222-280 AD) in the Yangtze River region, the Pearl River region, and North Vietnam. Cao Cao (155-220 AD) established the foundation for *Cao Wei* and controlled Emperor Xian of Han in 196-220 AD. In 220 AD, Cao Pi, Cao Cao's son, forced Emperor Xian of Han to step down and established *Cao Wei*. *Shu-Han* was established by Liu Bei (161-223 AD, as emperor in 221-223 AD). In 263AD, Liu Shan (207-271 AD, as emperor in 223-263 AD), Liu Bei's son, surrendered to *Cao Wei*. In 265 AD, Cao Huan (246-302 AD, as emperor in 260-265 AD), the emperor of *Cao Wei*, was forced to step down by Sima Yan (236-290 AD, as emperor in 266-290 AD) who established Jin Dynasty and was known as Emperor Wu of Jin. *Eastern Wu* was established by Sun Quan (182-252 AD, as emperor in 222-252 AD). In 280 AD, Sun Hao (243-284 AD, as emperor in 264-280 AD), the emperor of *Eastern Wu* and the grandson of Sun Quan, surrendered to Sima Yan and ended the Three Kingdoms period; at the end of the last chapter of *Romance of the Three Kingdoms*, there was a comment on this event:"The major trend of the world (*Tien Xia*) is division after long period of unification and unification after long period of division", see Fig. 5(b).

It seems that what Luo Guanzhong had in mind about *Tien Xia* included only the regions of the Yellow River and the Yangtze River; he wrote that Emperor Gaozu of Han unified the world as shown in Fig. 5(a), but Emperor Gaozu of Han never took Nanyue in the Pearl River region and North Vietnam into the territory of the Han Dynasty while Nanyue belonged to the territory of Qin Dynasty (221-206BC). Zhao Tuo (about 240 BC to 137 BC) was a general of Qin Dynasty. After the collapse of the Qin Dynasty, Zhao Tuo established Nanyue as an independent state with the capital at Panyu (now Guangzhou).

Book of Han by Ban Gu is a book about Han Dynasty. Figures 5(c) and 5(d) contain the historical records that Emperor Wu of Han took Nanyue (including Jiaozhi, Jiuzhen, and Rinan in the middle and north Vietnam) and Korea (including Zhenfan, Lintun, Lelang and Xuantu 4 Commanderies) into the territory of Han Dynasty in 111BC and 108 BC, respectively. The text in Figures 5(b) indicated that Luo Guanzhong considered that Emperor Wu of Jin had unified *Tien Xia* in 280AD, but in this year the original territories of Zhenfan, Lintun, Lelang and Xuantu did not belong to Jin Dynasty. During the life time of Luo Guanzhong, the original territories of Jiaozhi, Jiuzhen, and Rinan did not belong to Yuan or Ming Dynasties. Thus Luo Guanzhong's statements about *Tien Xia* in Figs. 5(a) and 5(b) could be supported by historical records when *Tien Xia* included only the regions of the Yellow River and the Yangtze River.

Dr. Monty G. Marshall, a Principal Investigator of Societal-Systems Research Inc., has executed a Polity IV Project with the website: http://www.systemicpeace.org/polity/polity4.htm, which contains data about Annual Polity scores (APS) from -10 to 10 for 167 countries in 1946-2013. The countries include following categories with related scores: Full democracy (10), Democracy (6 to 9), Open Anocracy (1 to 5), Closed Anocracy (-5 to 0), Autocracy (-10 to -6), Failed or Occupied. From the data in this website, one can easily find that from 1980 to 1990 to 2010, Annual Polity scores of most countries increase or maintain at the highest value 10. Figure in the web-site http://www.systemicpeace.org/polity/polity1.htm shows that the number of democracy countries increases and the number of autocracy countries decreases after 1980.

The Annual Polity scores of Taiwan can be found in Fig. 6. Here I would like to explain the data in Fig. 6 and present a simple model for the evolution pattern.

According to the Republic of China (ROC) Constitution passed by National Assembly on 25 December 1946 in Nanjing and executed from 25 December 1947, National Assembly is responsible for election of ROC President and Vice President, and the modification of the Constitution. Executive Yuan is responsible for administration of ROC Central Government. Legislative Yuan is responsible for passing laws and government budgets. Members of National Assembly and Legislative Yuan should be re-elected every 6 and 3 years, respectively. A person can be ROC President for consecutive two terms and each term has 6 years. Chiang Kai-shek (31 October 1887-5 April 1975) was elected as ROC President on 20 April 1948 in Nanjing.

KMT (Kuomingtang, China National Party) failed in the war with China Communist Party in China and moved ROC Central Government, National Assembly, Legislative Yuan, etc, from Nanjing to Taipei in 1949. Thus the data in Fig. 6 started in 1949. Since members of the National Assembly and Legislative Yuan were moved from China to Taiwan and they were not re-elected, Taiwan Annual Polity scores from 1949 to 1974 were at very low level of -8.

The imposition of Taiwan martial law started on 20 May 1949, and the publication of new newspapers and the formation of new political parties were prohibited. When Wu Kuo-Cheng (1903-1984) was Taiwan governor in 1949-1953, Taiwan started elections of city mayors, county heads, and councillors via popular votes in 1950-1951. Most elected persons were KMT members, but there were still some non KMT members, e.g. Wu San-Lien (1899-1988) was elected as Taipei mayor on 7 January 1951, Henry Kao (1913-2005) was elected as Taipei mayor on 2 May 1954 and Yu Deng-fa (1904-1989) was elected as Kaohsiung county head on 24 April 1960.

Chiang Kai-shek was elected as ROC President on 21 March 1954, 21 March 1960, 21 March 1966, and 21 March 1972 in Taipei by the National Assembly. In the 1966 and 1972 elections, Yen Chia-kan (1905-1993) was elected as the Vice President. In 1960, National Assembly managed to modify the ROC Constitution so that Chiang Kai-shek could be elected as ROC President for the third term, the 4th. term, etc. There was a movement against the modification of the ROC constitution and Chiang Kai-sheks re-election in 1960 led by Lei Chen (1897-1979) who was the key responsible person of the periodical Free China Journal (1949-4 September 1960). After the ROC constitution was modified and Chiang Kai-shek's election into the third term in 1960, there was a movement to form a new party: China Democratic Party (CDP) led by Lei Chen. This movement was suppressed by KMT by putting Lei Chen and his supporter Fu Chen (1927-1991) in jail on 4 September 1960. Lei Chen was in jail for 10 years and Fu Chen lost his freedom for 6 years. Free China Journal was also terminated. Other participants of the democratic movement in 1960 did not continue the movement and KMT thus successfully suppressed the democratic movement in 1960.

From 1969, Taiwan citizens began to elect a small number of persons to join the original large numbers of members in National Assembly and Legislative Yuan, e.g. Huang Hsin-chieh (1928-1999) and Kang Ning-hsiang (born 1938) were first elected as members of Legislative Yuan in 1969 and 1972, respectively, and thus Annual Polity scores had slightly increased from 1976 to 1986. There were bigger jumps after 1986. Such bigger jumps could be related to democratic movement after 1977.

In 1971, ROC government in Taipei lost the seat in United Nations (UN), more and more people, including students, paid attention to the revision to the political system. In May 1972, Chiang Ching-kuo (27 April 1910-13 January 1988) was appointed by his father Chiang Kai-shek as the President of the Executive Yuan. Chiang Ching-kuo paid more attention to economic development in Taiwan and had a stronger control over Taiwan society. On 23 December 1972, all elected mayors and county heads were KMT members, Hsu Hsin-liang (born 27 May 1941), also a KMT member, was elected as a councillor of Taiwan Provincial Council.

On 5 April 1975, Chiang Kai-shek passed away and Yen Chia-kan became the ROC President; punishments for all prisoners were reduced and Shih Ming-teh (born 15 January 1941, as a political prisoner since 1962) was released on 16 June 1977. On 19 November 1977, Taiwan had election of city mayors, county heads, and councillors. Huang Hsin-chieh and Kang Ning-hsiang organized collective champion activities to support non-KMT candidates, called Tangwai, meaning outside KMT. Hsu Hsin-liang was not nominated by KMT and run for Taoyuan county head as an independent candidate. A voter in Zhongli city of Taoyuan county reported to witness the KMT rigging the election, which leaded to Zhongli incident, that made electoral fraud greatly reduced in all Taiwan in this election. Among 20 seats of mayors and county heads, KMT lost 4 seats including Taoyuan county; among 77 seats of councillors of Taiwan Provincial Council, KMT lost 21 positions. The later included Lin Yi-hsiung (24 August 1941) and Chang Chun-hung (born 17 May 1938).

On 20 May 1978, Chiang Ching-kuo took the position of ROC President. On 23 December 1978, there were planed elections of additional members for National Assembly and Legislative Yuan. The election was canceled due to the announced US-PRC diplomatic relation starting on 1 January 1979. In January 1979, Yu Deng-fa invited Tangwai to his home town for a meeting on 1 February 1979 to exchange ideas about what to do in the next step. Before the meeting, Yu Deng-fa and his son were arrested. Hsu Hsin-liang, Shih Ming-teh, et al. organized a protest demonstration in Yu Deng-fa's home town Qiaotou. This Qiaotou incident lead KMT to freeze Hsu Hsin-liang's position of Taoyuan county head for two years, Hsu then went to USA.

Huang Hsin-chieh, Shih Ming-teh, Lin Yi-hsiung, Chang Chun-hung, Yao Chia-wen (born 15 June 1938), Lu Annette (born 6 June 1944), Chen Chu (born 10 June 1950), Lin Hung-hsuan (1942-2015), et al. continued the democratic movement and published the political magazine "Formosa". Some leaders of the magazine organized a march in Kaohsiung to celebrate the world human-right day on 10 December 1979. The activity was constrained by policemen and military personnel, which caused the conflict between two sides and leaded to Formosa Incident, also known as Kaohsiung Incident [61]. Most active members of the Formosa magazine were arrested from 13 December 1979 and Formosa magazine was terminated. Shih Ming-teh escaped with the help from the Presbyterian Church and was arrested on 8 January 2000. Lawyers, e.g. Chang Teh-ming (born 1938), Chen Shui-bian (born 12 October 1950), Frank Hsieh (19 May 1946), Su Tseng-chang (28 July 1947), et al. helped arrested persons in the court. Shih Ming-teh, Huang Hsin-chieh, Lin Yi-hsiung, Chang Chun-hung, Yao Chia-wen, Lu Annette, Chen Chu, and Lin Hung-hsuan were accused in the military course for overthrowing the government. On 28 February 1980, the first day of the court process, Lin Yi-hsiung's mother and twin daughters were killed to death in the basement of Lin's home. Lin's elder daughter was also killed, but found earlier and sent to a hospital to rescue, and is still alive. On 18 April 1980, Shih Ming-teh and Huang Hsin-chieh were sentenced to life in prison and 14 years in prison, respectively; other 6 persons were sentenced to 12 years in prison [61]. In April-May 1980, another group of 33 people were tried in civil court and sentenced to terms ranging from two to six years [61]. The third group of persons helped Shih Ming-teh to escape; this group included Kao Chun-ming (born 6 June 1929) and Chang Wen-ing (born 26 July 1950) who were sentenced to 7 and 4 years in prison, respectively.

On 6 December 1980, the supplementary elections for the National Assembly and the Legislative Yuan canceled in December 1978 were restored. Among 70 directly elected members of Legislative Yuan, Tangwai took 9 seats, including Kang Ning-hsiang, Chang Teh-ming, Huang Tien-fu (born 1938, Huang Huang Hsin-chieh's younger brother) and Hsu Jung-shu (27 December 1939, Chang Chun-hung's wife). Among 76 elected members of National Assembly, Tangwai took 14 seats, including Chou Ching-yu (born 12 June 1944, Yao Chia-wen's wife) who got the largest number of votes in all Taiwan. Thus Tangwai had slightly recovered from the great damage of Formosa Incident [61].

On 20 May 1984, Chiang Ching-kuo took the position of ROC President for the second term, and Lee Teng-hui (born 15 January 1923) [62] took the position of ROC Vice President.

On 15 October 1984, Henry Liu (1932-15 October 1984) was murdered in the garage of his home in Daly City, California [63]. It was found that ROC Military Intelligence Bureau, Vice Admiral Wang Hsi-ling ordered Chen Chi-li, a gangster head, to do so. This event greatly damaged the image of KMT government.

On 28 September 1986, Tangwai announced to establish a new political party: Democratic Progressive Party (DDP) in a meeting in the Grand Hotel. This time KMT did not try to suppress Tangwai as in 1979. On 7 October 1986, Chiang Ching-kuo told Katherine C. Graham of Washington Post that his government would release the martial law, allow the publication of new newspapers, and formation of new political parties. In the election on 6 December 1986, among 55 regional seats of elected new members of the Legislative Yuan, DPP took 11 seats; among 84 seats of the National Assembly, DDP took about 15 seats.

On 15 July 1987, Taiwan martial law was lifted. On 13 January 1988, Chiang Ching-kuo passed away, and Lee Teng-hui became ROC (Taiwan) president. In March 1990, National Assembly had a meeting in Taipei to elect ROC president and Vice President. There were 752 members and only 84 persons were elected in Taiwan on 6 December 1986, and other members were elected before 1949. Such a situation inspired a student movement, called Wild-Lily student movement [64] starting on 16 March 1990 at the Chiang Kai-Shek (CKS) Memorial Hall. On 18 March 2018, the representatives of students made their four demands [64]: 1. Abolish the National Assembly and re-establish a new National Assembly infrastructure, 2. Nullify the Temporary Provisions Against the Communist Rebellion and re-establish constitutional order, 3. Hold a National Affairs Conference, 4. Establish a political reform time table.

On 20 March 1990, there were more than 5,000 protestors. On 21 March 1990, Lee Teng-hui was elected as ROC president. After the election, President Lee invites 50 students to the Presidential Hall and gave positive response to the students' request. The students ended the movement on 22 March 1990. In May 1990, President Lee announced the invalidation of trials for the Formosa Incident [61]. In June 1990, Justices of Constitutional Court of the Judicial Yuan made Explanation 261: All members of National Assembly and Legislative Yuan elected before 1949 should step down before 31 December 1991.

On 21 December 1991, there were elections of 325 members of the Second National Assembly. In this election, KMT took 254 seats, DDP took 66 seats, other parties and independent persons took 5 seats. All members of National Assembly and Legislative Yuan elected before 1949 had stepped down before 31 December 1991. From 1 January 1992, National Assembly and Legislative Yuan operated by representatives elected in Taiwan and Annual Polity Scores of Taiwan in Fig 6 had reached 7 and Taiwan became a democratic country. On 19 December 1992, there were elections of 161 members of the Second Legislative Yuan, in this election KMT took 95 seats, DDP took 51 seats, other parties and independent candidates took 15 seats.

In July 1994, the Second National Assembly revised the ROC Constitution, so that ROC President and Vice President should be elected directly by citizens in Taiwan region via a popular vote. On 23 March 1996, Taiwan citizens elected Lee Teng-hui as ROC (Taiwan) President and Lien Chan (born 27 August 1936) as ROC (Taiwan) Vice President.

On 18 March 2000, Taiwan citizens elected DDP members Chen Shui-bian as ROC (Taiwan) President and Lu Annette as ROC (Taiwan) Vice President. This was the first time that DDP won the presidential election, and there was a peaceful transfer of power from one party to another; the Annual Polity Score of Taiwan in Fig 6 had reached 9. On 20 March 2004, Taiwan citizens again elected Chen Shui-bian as ROC (Taiwan) President and Lu Annette as ROC (Taiwan) Vice President and the Annual Polity Score of Taiwan in Fig 6 had reached the highest value 10. From 20 May 2000 to 20 May 2008, DDP took the President Office, but the number of seats DDP had in the Legislative Yuan was smaller than 50%. From 20 May 2008 to 20 May 2016, KMT took the President Office and controlled the Legislative Yuan. On 16 January 2016, DDP won both President election and election of the Legislative Yuan. From 20 May 2016, DDP took the President Office and controlled the Legislative Yuan and thus could passed some laws so that political parties in Taiwan can compete on more equal basis. For the election on 16 January 2016, KMT donated KMT president candidate more than 2×10^8 new

Taiwan dollars, and DDP donated DDP president candidate less than 10^7 new Taiwan dollars.

Here I would like to present a simple model for the transition of Taiwan from an autocracy country in 1949-1975 to a full democracy country after 2004 as shown in Fig 6 by using the analogy of such a transition with behavior of the gas-liquid system shown in Figure 1(a).

Figure 1(a) is a phase diagram for gas-liquid system of CO_2. A′CE′ curve is an isothermal curve passing the critical point C with the critical temperature $T_c = 31.3°C$. A-B-C-D-E is an isothermal curve for the temperature $T = 21.5°C$. below the critical temperature T_c. A to B is in the liquid phase. When there is a fluctuation, e.g. due to cosmic ray radiation, at the point between A and B, the fluctuation can be suppressed easily. When there is no impurity in the system and the volume increases, the system can maintain at the liquid state and reach point b, which is a meta-stable state. When there is a fluctuation, the system at point b can decay easily and reach a stable state between B and D, which is a mixture of liquid and gas.

In 1949-1975, Taiwan was an autocracy country and was similar to liquid state in Fig. 1(a). The democratic movement lead by Lei Chen in 1960 was similar to a fluctuation at a point between A and B in Fig. 1(a) and it was suppressed by KMT easily. Due to economic growth, more and more people received higher education, etc, but KMT still controlled the society tightly via Taiwan martial law, Taiwan reached a meta-stable state in 1976-1985, similar to point b in Fig. 1(a); Yu Teng-fa incident and Formosa incident [61] in 1979 shown in Fig 6 were similar to fluctuations near point b in Fig. 1(a). KMT could not suppress such fluctuations and Taiwan was approaching to a two-parties political system after 1986, similar to a point in the horizontal line BC in Fig. 1(a), which is a mixture state.

It seems that the model for the increase of annual polity scores of Taiwan can be applied to other countries, thus the number of democracy countries increases and the number of autocracy countries decreases after 1980 as shown in http://www.systemicpeace.org/polity/polity1.htm. The time scale for such increase might depend on the system size. For a large country, it might take longer time to transfer from an autocracy country to a democracy country.

5. Summary and discussion

We have shown that concepts and methods of statistical physics could be useful for understanding universal and scaling behaviors in complex many-body systems, including human and social systems. We are using such concepts and methods to study evolution of root networks of Chinese characters [65] and universal behaviors and patterns in Chinese history.

Results of Section 3 inspire other problems for further studied as discussed in [1].

There exists other democracy index, e.g. https://en.wikipedia.org/wiki/Democracy_Index, but data for such index do not cover many years as Fig. 6 for studying evolution pattern.

6. Acknowledgments

The author thanks collaborators, R. E. Amritkar, Wen-Jong Ma, Shih-Chieh Wang, Chi-Ning Chen, Wei-Cheng Kuo, for the work reported in paper. He thanks Armen Allahverdyan, Monty G. Marshall, Johannes Voit, and Karen Petrosyan for critical reading of the paper. He also thanks Deng Weibing and Wei Li for inviting him to give a talk at The 5th International Workshop on Statistical Physics and Mathematics for Complex Systems (SPMCS2017), in Wuhan, China on 11-15 October 2017; this paper was based on this talk. This work was supported by Grant MOST 106-2112-M-001-027.

References

[1] Hu CK and Kuo WC 2005 in *POLA Forever: Festschrift in honor of Prof. W. S. Y. Wang on his 70th birthday* Dah-An Ho and Ovid J. L. Tzeng ed. (Inst. of Linguistics, Acad. Sin., Taipei) ISBN: 986-00-3151-7

[2] Stanley HE 1971 *Introduction to Phase Transitions and Critical Phenomena* (Oxford Univ. Press, New York)

[3] Hu CK 2014 *Chinese J. Phys.* **52** 1-76.

[4] Andrews T 1869 *Philosophical Transactions of the Royal Society of London* **159** 575
[5] Brush SG 1976 *The Kind of Motion We Call Heat 1* (North-Holland, Amsterdam)
[6] van der Waals JD 1873 *Over de Continuiteit van den Gas- en Vloeistoftoestand (on the continuity of the gas and liquid state)* PhD thesis (excerpt) (Leiden, The Netherlands)
[7] Hu CK 1986 *Science Monthly (Taipei)* **17** 170 (in Chinese)
[8] Jones JE 1924 *Proc. R. Soc. London A* **106** 463
[9] Lennard-Jones JE 1931 *The Proceedings of the Physical Society* **43** part 5
[10] Guggenheim EA 1945 *J. Chem. Phys.* **13** 253
[11] Yang CN and Lee TD 1952 *Phys. Rev.* **87** 404; Lee TD and Yang CN 1952 *Phys. Rev.* **87** 410
[12] Ising E 1925 *Zeitschrift fur Physik* **31** 253
[13] Wilson KG 1971 *Phys. Rev. B* **4** 3174 and 3184; Wilson KG 1972 *Phys. Rev. Lett.* **28** 548
[14] Wilson KG and Fisher ME 1972 *Phys. Rev. Lett.* **28** 240
[15] Wilson KG 1983 *Rev. Mod. Phys.* **55** 583
[16] Blöte HWJ, Luijten E and Heringa JR 1995 *J. Phys. A: Math. Gen.* **28** 6289-6313
[17] Talapov AL and Blöte HWJ 1996 *J. Phys. A: Math. Gen.* **29** 5727
[18] Sengers JV and Shanks JG 2009 *J. Stat. Phys.* **137** 857
[19] Watanabe H, Ito N, and Hu CK 2012 *J. Chem. Phys.* **136** 204102
[20] Metropolis N and Ulam S 1949 *Journal of American Statistical Association* **44** 335
[21] Metropolis N, Rosenbluth AW, Rosenbluth MN, Teller AH and Teller E 1953 *J. Chem. Phys.* **21** 1087
[22] Hu CK 1992 *Phys. Rev. B* **46** 6592-6595; Hu CK 1992 *Phys. Rev. Lett.* **69** 2739-2742
[23] Wu MC and Hu CK 2002 *J. Phys. A.* **35** 5189-5206
[24] Ivashkevich EV, Izmailian NS and Hu CK 2002 *J. Phys. A.* **35** 5543-5561
[25] Privman V and Fisher ME 1984 *Phys. Rev. E* **30** 322
[26] Hu CK 1994 *J. Phys. A* **27** L813-L820
[27] Hu CK, Lin CY and Chen JA 1995 *Phys. Rev. Lett.* **75** 193-196; 1995 *Physica A* **221** 80-88
[28] Hu CK and Lin CY 1996 *Phys. Rev. Lett.* **77** 8-11
[29] Wang FG and Hu CK 1997 *Phys. Rev. E* **56** 2310-2313
[30] Lin CY and Hu CK 1998 *Phys. Rev. E* **58** 1521-1527
[31] Okabe Y, Kaneda K, Kikuchi M and Hu CK 1999 *Phys. Rev. E* **59** 1585-1588
[32] Tomita Y, Okabe Y and Hu CK 1999 *Phys. Rev. E* **60** 2716-2720
[33] Izmailian NS and Hu CK 2001 *Phys. Rev. Lett.* **86** 5160-5163
[34] Wu MC, Hu CK and Izmailian NS 2003 *Phys. Rev. E* **67** 065103
[35] Li MS, Kouza M and Hu CK 2007 *Biophysical J.* **92** 547-561
[36] Hu CK 2013 *AIP Conf. Proc.* **1518** 541
[37] Wu MC, Li MS, Ma WJ, Kouza M and Hu CK 2011 *EPL* **96** 68005
[38] Gevorkian SG, Allahverdyan AE, Gevorgyan DS and Hu CK 2011 *EPL* **95** 23001
[39] Gevorkian SG, Allahverdyan AE, Gevorgyan DS, Simonian AL and Hu CK 2013 *Plos One* **8** e78526
[40] Chiang HL, Chen CJ, Okumura H and Hu CK 2014 *J. Comp. Chem.* **35** 1430-1437
[41] Hu CK 2015 *Journal of Physics: Conference Series* **604** 012009
[42] Gevorkian SG, Allahverdyan AE, Gevorgyan DS and Hu CK 2015 *Sci. Rep.* **5** 13064
[43] Ma WJ and Hu CK 2017 *Sci. Rep.* **7** 3105
[44] Brown R 1828 *Edinburgh New Philosophical Journal* July-September 358-371
[45] Bachelier L 1900 *Ann. Sci. Ec. Normale Super.* **17** 21
[46] Einstein A 1956 *Investigation on the Theory of the Brownian Movement* (Dover, New York)
[47] Langevin L 1908 *C. R. Acad. Sci. (Paris) 146* 530-533
[48] Lemons DS and Gythiel A 1997 *Am. J. Phys.* **65** 1079-1081
[49] Ma WJ, Wang SC, Chen CN and Hu CK 2013 *EPL* **102** 66003
[50] Mantegna BN and Stanley HE 1995 *Nature* **376** 46-49
[51] Saakian DB, Martirosyan A, Hu CK and Struzik ZR 2011 *EPL* **95** 28007
[52] Laloux L, Cizeau P, Bouchaud JP and Potters M 1999 *Phys. Rev. Lett.* **83** 1467
[53] Plerou V, *et al.* 1999 *Phys. Rev. Lett.* **83** 1471
[54] Plerou V,*et al.* 2002 *Phys. Rev. E* **65** 66126
[55] Ma WJ, Hu CK and Amritkar RE 2004 *Phys. Rev. E* **70** 26101
[56] See also Noh JD 2000 *Phys. Rev. E* **61** 5981
[57] Reichl LE 2009 *A Modern Course in Statistical Physics*, 3rd ed., (Weily-VCH, Weinheim)
[58] Zipf GK 1949 *Human Behavior and the Principle of the Least Effort* (Addison-Wesley, Reading, MA)
[59] Yang ACC, Peng CK, Yien HW and Goldberger AL 2003 *Physica A* **329** 473-483
[60] Chi TT, Wang SC and Hu CK 2015 *Physica A* **437** 408-417
[61] https://en.wikipedia.org/wiki/Kaohsiung_Incident.

IOP Conf. Series: Journal of Physics: Conf. Series **1113** (2018) 012002 doi:10.1088/1742-6596/1113/1/012002

[62] https://www.britannica.com/biography/Lee-Teng-hui.
[63] https://en.wikipedia.org/wiki/Henry_Liu.
[64] https://outreachfortaiwan.org/social-movements/wild-lily-student-movement/.
[65] Lee PH, Chen JL, Shen FH, Wang PC, Chi TT, Xiao ZR, Jhang ZJ, Yeh YN, Chen YY and Hu CK 2018 preprint.

A true least action principle for damped motion

Qiuping A. Wang[1,2] and Ru Wang[2]
[1]ESIEA GROUP, 9 rue Vesale, 75005 Paris, France
[2]IMMM, UMR CNRS 6283, Le Mans University, 72085 Le Mans, France

E-mail: alexandre.wang@univ-lemans.fr

Abstract. This work is a formulation of the least action principle for classical mechanical dissipative systems. We consider a whole conservative system composed of a damped moving body and its environment receiving the dissipated energy. This composite system has a conservative Hamiltonian $H = K_1 + V_1 + H_2$ where K_1 is the kinetic energy of the moving body, V_1 its potential energy and H_2 the energy of the environment. The Lagrangian is found to be $L = K_1 - V_1 - E_d$ where E_d is the energy dissipated from the moving body into the environment. The usual variation calculus of least action leads to the correct equation of the damped motion.

1. Introduction

The Least Action Principle (LAP) is one of the most valuable heritages from the classical mechanics[1-3]. The fact that the formulation of the whole classical physics as well as of the quantum theory in its path integral formalism[4]could be based on or related to this single mathematical rule gives to LAP a fundamental priority to all other visibly different principles, empirical laws and differential equations in different branches of physics. This priority of LAP has nourished two ambitions of physicists. The first one is the (rather controversial) effort to deepen the understanding of nature through this principle and to search for the fundamental meaning of its exceptional universality in physics[3,5,6]. The second one is to extend it to more domains such as thermodynamics and statistical mechanics (with the pioneer efforts of Boltzmann, Helmholtz and Hertz[7]), stochastic dynamics (e.g., large deviation theory[8] and stochastic mechanics[9]), and dissipative mechanical systems[10-12]. The present work follows this last effort to formulate LAP for classical mechanical dissipative systems.

LAP was originally formulated only for Hamiltonian system[2], i.e., the sum $H = K + V$ of kinetic energy K and potential energy V of the considered system satisfies the Hamiltonian equations. For Hamiltonian systems, any real trajectory between two given configuration points must satisfy the LAP, a vanishing first variation of the action A created by tiny deformation of the trajectory[2,6]:

$$\delta A = \delta \int_0^{t_b} Ldt = \int_0^{t_b} \delta Ldt = 0 \tag{1}$$

where the action $A = \int_0^{t_b} Ldt$ is a time integral of the Lagrangian $L = K - V$ on the trajectory over a fixed time period t_b. One of the important results of this variational calculus is the

Euler-Lagrange equation given by[6] (for one freedom x)

$$\frac{d}{dt}\left(\frac{\partial L}{\partial \dot{x}}\right) - \frac{\partial L}{\partial x} = 0 \tag{2}$$

where $\dot{x}$ is the velocity. In many cases when H and L do not depend on time explicitly, a Hamiltonian system is energy conservative.

A problem of variation takes place for nonconservative systems whose dissipative force, say, f_d is artificially introduced into the equation of motion in this way $\frac{d}{dt}\left(\frac{\partial L}{\partial \dot{x}}\right) - \frac{\partial L}{\partial x} = f_d$. This is equivalent to write $\int_0^{t_b}(\delta L + f_d \delta x)dt = 0$. However, it is impossible to define an action integral with a single (Lagrangian) function satisfying Eq.(2). Hence LAP is absent for dissipative systems.

There has been a longstanding effort to recover LAP for dissipative system[11]. Since all systems in nature are energy dissipative, this formulation is essential for this beautiful fundamental principle to be better understood, applied to a much wider range of systems and to be related to many other scientific principles relative to the energy dissipation[10,11,13].

As far as we know, the first variational calculus applied to damped motion dates back to Euler's work in 1744 for the brachistochrone (shortest time) problem with friction[13]. More recently, Rayleigh[15] has introduced a 'dissipative function' $D = \frac{1}{2}\zeta\dot{x}^2$, for the special case of the Stokes' law $f_d = -m\zeta\dot{x}$, to write $\frac{d}{dt}\left(\frac{\partial L}{\partial \dot{x}}\right) + \frac{\partial D}{\partial \dot{x}} - \frac{\partial L}{\partial x} = 0$, where ζ is the drag constant and m the mass of the damped body. Nevertheless, LAP is not recovered since there is no Lagrangian for defining an action which satisfies Eq.(1). Other major propositions include the Bateman approach[16] to introduce complementary variables and equations, the definition of dissipative Lagrangian by multiplying the non dissipative one with an exponential factor $exp(\zeta t)$[17] where t is the time, the fractional derivative formulation[18], and the pseudo-Hamiltonian mechanics[19] where a parameter was introduced to characterize the degree of dissipation. The reader is referred to the reviews in [10,11,14,18,19] about the details of these propositions. In general, the Lagrangian in these solutions is not unique and has no energy connection like $L = K - V$ (see for instance the quasi-Lagrangian $L = e^{\zeta t}(K - V)$ and the corresponding quasi-Hamiltonian $H = e^{-\zeta t}K + e^{\zeta t}V$ for damped harmonic oscillator[17]). Hence no variational or optimal calculus was possible in general form[10,11,14].

A common character of these previous works is to consider the damped body as an isolated body in the calculations. The problem is that a dissipative system is coupled to an environment and loses energy into the latter which becomes an integral part of the motion. As far as this lost energy is not considered, the Lagrangian of the 'isolated' body inevitably loses energy connection and generic optimal characters[10,11] as mentioned above.

The aim of this work is to recover LAP in a general way for dissipative system with an energy connected and unique Lagrangian which is related as usual to the conservative Hamiltonian by Legendre transformation. The three conventional formulations of analytical mechanics, i.e., the Hamiltonian, Lagrangian and the Hamilton-Jacobi equations, are all preserved.

2. The conservative Hamiltonian

Our basic idea is to consider the damped moving body (system 1) and its environment (system 2) as a whole conservative system. The total Hamiltonian includes the instantaneous kinetic and the potential energy of the body, as well as the mechanical energy transformed into heat or other forms of energy (noises, vibration, electromagnetic radiation etc.) into the environment. The body moves along the axis x with velocity $\dot{x}$. Its environment is composed of N bodies with position x_i and velocity $\dot{x}_i$ $(i = 1, 2, ..., N)$. The energy transfer from system 1 to system 2 occurs only through a friction force. The total Hamiltonian reads $H = K_1 + V_1 + K_2 + V_2 + H_{int}$ where $K_1 = \frac{1}{2}m\dot{x}^2$ is the kinetic energy and V_1 the potential energy of the system 1,

$K_2 = \frac{1}{2}\sum_i^N m_i\dot{x}_i(t)^2$ the kinetic energy and $V_2 = \sum_i^N v(x_i)$ the potential energy of system 2 with $v(x_i)$ the potential energy of the particle i, and H_{int} the interaction energy between the system 1 and 2. H_{int} is responsible for the friction law and determined by the coupling mechanism on the interface between the moving body and the environment. We can suppose that, for a limited time period of the motion under consideration, this interface (body's shape and size, body-environment distance, nature of the closest parts of the environment to the interface, etc.) and the friction law do not change significantly, hence the interaction mechanism should not change with the virtual variation of paths. In this case H_{int} can be neglected in the variational calculus and we only have $H = K_1 + K_2 + V_1 + V_2$ or $H = H_1 + H_2$ where $H_1 = K_1 + V_1$ is the total energy of the system 1 and $H_2 = K_2 + V_2$ that of the system 2.

It should be stressed that the impact of the thermal fluctuation in system 2 on system 1 should be neglected. The motion of the system 1 remains classical mechanical and deterministic. This is reasonable for a mechanical body which is much larger than the constituents of system 2 and has much larger energy variation during the motion than the energy fluctuation of the thermal motion in system 2.

3. Variational calculus of LAP for damped motion

The Lagrangian L of the whole conservative system can be defined by using the Legendre transformation $L = p\dot{x} + \sum_i^N p_i(t)\dot{x}_i(t) - H = K_1 - V_1 + 2K_2 - H_2$ where p is the momentum of system 1 and p_i the momentum of the particle i of the system 2. The corresponding action is given by

$$A = \int_0^{t_b} Ldt = \int_0^{t_b} (K_1 - V_1 + 2K_2 - H_2)dt. \tag{3}$$

H_2 can be expressed as a function of $x_i(t)$ and the velocity $\dot{x}_i(t)$ of the constituent parts of the system 2. Its general expression reads

$$H_2 = \sum_i^N \frac{1}{2} m\dot{x}_i^2(t) + V_2[x_1(t), x_2(t)...x_N(t)] \tag{4}$$

With this expression of H_2, the Lagrangian only depends on its variables at the time moment t, i.e.,

$$L = K_1(\dot{x}(t)) - V_1(x(t)) + 2K_2(\dot{x}_i(t)) - H_2(x_i(t), \dot{x}_i(t)) \quad i = 1, 2, ..., N \tag{5}$$

The usual variation calculus with a tiny variation $\delta x(t)$ of the path $x(t)$ damped body yields :

$$\begin{aligned} \delta A &= \int_0^{t_b} [\frac{\partial L}{\partial x}\delta x + \frac{\partial L}{\partial \dot{x}}\delta \dot{x}]dt \\ &= \int_0^{t_b} [\frac{\partial L}{\partial x} - \frac{d}{dt}\frac{\partial L}{\partial \dot{x}}]\delta x dt. \end{aligned} \tag{6}$$

where we have made a time integration by parts of $\delta\dot{x}$ with the conditions $\delta x(a) = \delta x(b) = 0$. The LAP requires $\delta A = 0$, which, due to the arbitrary nature of $\delta x(t)$, leads to

$$\frac{\partial L}{\partial x} - \frac{d}{dt}\frac{\partial L}{\partial \dot{x}} = 0. \tag{7}$$

This is the differential equation for the damped motion of the system 1.

In order to make calculation with Eq.(7), H_2 must be written as an explicit function of x and $\dot{x}$. For this purpose, we suppose that H_2 changes in time only due to the energy dissipation of the system 1. We can write $H_2(t) = H_2(t_a) + E_d(x(t))$ where $H_2(t_a)$ is the energy of the system

2 at the initial time t_a (a constant for the motion hereafter), and $E_d(x(t))$ the energy received from system 1 through the work of the friction force $f_d(x,\dot{x})$ from the initial moment t_a to a moment t $(0 \leq t \leq t_b)$:

$$E_d(x(t)) = -\int_{x_a}^{x(t)} f_d(x(\tau),\dot{x}(\tau))dx(\tau) \tag{8}$$

where τ is any time moment between $t_a = 0$ and t, and $dx(\tau)$ a small displacement along the path at the time τ. The Lagrangian can be written as $L(x(t),\dot{x}(t)) = K_1(\dot{x}(t)) + 2K_2(\dot{x}_i(t)) - V_1(x(t)) - E_d(x(t))$ where the constant $H_2(t_a)$ is dropped. Introducing this Lagrangian into Eq.(7) and considering the fact that K_1 and K_2 are not explicit function of $x(t)$ and that K_2, V_1 and E_d are not explicit function of $\dot{x}(t)$ at time t, one gets

$$\frac{d}{dt}\frac{\partial K_1}{\partial \dot{x}} + \frac{\partial V_1}{\partial x} + \frac{\partial E_d}{\partial x} = 0. \tag{9}$$

Considering $f_d(x(t),\dot{x}(t)) = -\frac{\partial E_d(x(t))}{\partial x(t)} = -\frac{\partial}{\partial x(t)}\int_0^{x(t)} f(x(\tau),\dot{x}(\tau))dx(\tau)$ [20], we straightforwardly obtain

$$m\ddot{x} = -\frac{\partial V_1}{\partial x} + f_d, \tag{10}$$

the Newtonian equation for the damped motion with the friction f_d.

To see the second formulism of classical mechanics, the Hamiltonian equations, we calculate the total time derivative of H reads $\frac{dH}{dt} = \frac{\partial H}{\partial t} + \frac{\partial H}{\partial x}\dot{x} + \frac{\partial H}{\partial p}\dot{p}$. Compare this to the same derivative made with the Legendre transformation $H = p\dot{x} - L$, i.e., $\frac{dH}{dt} = \dot{p}\dot{x} + \dot{p}\ddot{x} - \frac{\partial L}{\partial t} - \frac{\partial L}{\partial x}\dot{x} - \frac{\partial L}{\partial \dot{x}}\ddot{x}$ and considering the Euler-Lagrange equation Eq.(7), we get, besides $\frac{\partial H}{\partial t} = \frac{\partial L}{\partial t}$, the following Hamiltonian equations [6] :

$$\dot{p} = -\frac{\partial H}{\partial x}, \dot{x} = \frac{\partial H}{\partial p} \tag{11}$$

from which the conservation of total energy H can be calculated [2].

To see the third formalism of classical mechanics, the Hamilton-Jacobi equation, we relax t_b in the time integral of action or consider the integral as indefinite, and compute $L = \frac{dA}{de} = \frac{\partial A}{\partial t} + \frac{\partial A}{\partial x}\dot{x}$. Thanks to Eq.(7) and the Legendre transformation, we can get $p = \frac{\partial A}{\partial t}$ and the Hamilton-Jacobi equation :

$$\frac{\partial A}{\partial t} + H = 0 \tag{12}$$

for the whole system.

4. Concluding remarks

In summary, we have proposed a formulation of the least action principle for damped motion with an universal, energy connected and unique Lagrangian, thanks to the model of a conservative system composed of the damped moving body and of its environment. We hope that these results are helpful for further study of the relations between the fundamental principles of Lagrangian/Hamiltonian mechanics and the variational principles relative to energy dissipation [23-27]. It is shown that the three conventional formulations of analytical mechanics, i.e., the Hamiltonian, Lagrangian and the Hamilton-Jacobi equations, are all preserved.

This work is presented in term of friction as the coupling between the moving body and its environment. But the formulation is not limited to friction. This is an advantage of the substitution of H_2 by E_d which is written as a function of the damped motion without involving the details of the motions of the tiny particles in the environment. Hence the coupling can be any mechanism depending on the motion and dissipating energy from it. For instance, the emission of electromagnetic wave or light into the void (environment) from an accelerated charged body, or emission of sound by a vibrating body.

Finally we would like to mention that, although the term "least action" is used here for historical reason, the stationarity $\delta A = 0$ is not necessarily a minimum. The nature of the stationarity (minimum, maximum or inflection) of A has been addressed in our recent work [22] by numerical simulation of damped motion and of variational analysis, in which the action A_{op} along the optimal path is calculated and compared to the actions of many paths around the optimal one. One of the conclusions is that the stationarity of A_{op} undergoes a transition from minimum to maximum when drag constant ζ and the dissipated energy increase. Further investigation is necessary to confirm this evolution of action stationarity.

References

[1] Maupertuis P L M, 1750 *Essai de cosmologie*(Berlin) ; 1744 *Accord de différentes lois de la nature qui avaient jusqu'ici paru incompatibles Mém. As. Sc. Paris* p. 417;1746 *Les lois de mouvement et du repos, déduites d'un principe de métaphysique Mém. Ac. Berlin* p. 267
[2] Arnold V I 1989 *Mathematical methods of classical mechanics*(New York: Springer-Verlag)
[3] Stoltzner M 2003 *Studies in History and Philosophy of Science Part B: Studies in History and Philosophy of Modern Physics*,**34** 285
[4] Feynman R P and Hibbs A R 1965 *Quantum mechanics and path integrals*(New York: McGraw-Hill Publishing Company)
[5] Mach E 1987 *La mécanique, exposé historique et critique de son développement*(Paris:Edition Jacques Gabay)
[6] Lanczos C 1986 *The variational principles of mechanics*(New York: Dover Publication)
[7] I. Stengers 2003 ?*Cosmopolitiques I, Ed. La d—couverte*(Paris)
[8] Freidlin M I and Wentzell A D 1984 *Random perturbation of dynamical systems*(New York: Springer-Verlag)
[9] Yasue K 1981 *J. Math. Phys.*, **22** 110
[10] Sieniutycz S and Farkas H 2005 *Variational and extremum principles in macroscopic systems*(Elsevier)
[11] Vujanovic B D and Jones S E 1989 *Variational methods in nonconservative phenomena*(New York: Academic Press Inc.)
[12] Herrera L, Nunez L, Patino A and Rago H 1986 *A variational principle and the classical and quantum mechanics of the damped harmonic oscillator*, **54** 273
[13] Goldstine H H 1980 *A history of the calculus of variations from the 17th through the 19th century*(New York: Springer-Verlag)
[14] Gray C G 2009 *Scolarpedia* **4** 8291
[15] Goldstein H *Classical Mechanic*(Addison-Wesley: Mass)
[16] Bateman H 1931 *Phy. Rev.* **38** 815
[17] Sanjuan 1995 *Journal of Sound and Vibration* **185** 734
[18] Riewe F 1997 *Phys. Rev. E* **55** 3581.
[19] Duffin R J1962 *Arch. Rat. Mech. Anal.* **9** 309
[20] Weisstein E W *Second Fundamental Theorem of Calculus, From MathWorld–A Wolfram Web Resource. http://mathworld.wolfram.comSecondFundamentalTheoremofCalculus.html*
[21] Alembert J D 1990 *Traité de dynamique* (Sceaux: Editions Jacques Cabay)
[22] Hyon Y, Kwak D and Liu C 2009 *Energetic variational approach in complex fluids: Maximum dissipation principle, Discrete And Continuous Dynamical Systems* **26** 1291-1304
[23] Onsager L 1931 *Phys. Rev.* **37** 405
[24] Eylink G L 2008 *Physica D* **14**
[25] Prados A, Lasanta A and Hurtado P I 2011 *Phys. Rev. lett.* **107** 140601
[26] Koide T and Kodama T 2012 *J. Phys. A: Math. Theor.***45** 255204

IOP Conf. Series: Journal of Physics: Conf. Series **1113** (2018) 012004 doi:10.1088/1742-6596/1113/1/012004

Dynamical fluctuations in temporal networks based on factorial moment approach

Liping Chi and Chunbin Yang

College of Physical Science and Technology, Central China Normal University, Wuhan 430079, China

E-mail: chilp@mail.ccnu.edu.cn

Abstract. The factorial moments analyses are performed to study the scaling properties of dynamical fluctuations in temporal networks based on empirical data sets. The intermittent behavior is not evident between normalized factorial moment F_q and bin number M over the whole time range. But the strict power-law dependence of F_q on F_2, $F_q \sim F_2^{\beta_q}$, indicates that the interaction has self-similarity structure in time interval and the fluctuations are not purely random but dynamical and correlated. We further find that β_q has a well-fitted power-law relationship with $q-1$. The exponents are very close to that for systems undergoing second-order phase transition in Ginzburg-Landau theory and describe the strength of intermittent behavior in the dynamical fluctuations.

1. Introduction

Interactions in complex systems are not static but change over time, which can be modelled in terms of temporal networks [1]. Temporal network consists of a set of contacts (n_i, n_j, t), emphasizing on the time *when* node i and j have a connection. The addition of time dimension provides a new sight into the framework of complex network theory. In temporal networks, both structural properties and spreading dynamics crucially depend on the time-ordering of links.

The research of temporal networks has attracted great attention and it mainly focuses on two major aspects from the point view of time dimension. One is corresponding to the strategy of time aggregation especially when the topological characteristics are more relevant than the temporal properties. The topological structure of temporal network is achieved through aggregating contacts over a certain time interval and the temporal network is then represented as a series of snapshots of static graphs. Consequently, many existing concepts and tools of static graphs can be adopted to analyze temporal networks, since it is usually easier to analyze static networks. For example, the degree of a node $k_i(t)$ is described as the number of links that it has to other nodes within the time window $[t, t+\Delta t]$. The error and attack strategies in static networks have been applied to evaluate the temporal vulnerability [2], and so on. In order to understand the structure of temporal networks, it plays a crucial role to choose an optimal time interval Δt to construct static graphs from temporal networks. Krings *et al.* studied the influences of time intervals when aggregating the mobile phone network over time [3]. Holme analyzed three ways of constructing static snapshots from temporal networks [4], but no candidate weighing out as a best choice. It is now still an open question on how to choose the time interval to represent temporal networks.

The other aspect is related to using dynamical processes to probe into the influence of time series on temporal network. We should take into account the time-ordering of each contact and the inter-event time between two consecutive contacts. The inter-event time distribution in temporal network follows a power-law, which is also called burstiness [5]. Although it is recognized that time-ordering and bursty characters have strong influences on the dynamical processes of temporal networks, numerous studies have appeared to arrive at contradictory results. Lambiotte *et al.* had stressed that time-ordering and burstiness of contacts were critical in spreading process, which leaded to slow down spreading [6]. In the work of Rocha *et al*, they concluded that temporal correlations accelerated outbreaks [7] in SI and SIR model. Miritello *et al.* demonstrated that bursts hindered propagation at large scales, but group conversations favored local rapid cascades [8].

Despite the promoting results in temporal networks, this field is still in its early stages about how temporal effect and topological structure interplay and hence affect the dynamical process. In this paper, based on empirical data sets, we will investigate the scaling properties of the dynamical fluctuations of contacts and nodes in temporal networks by means of the factorial moment approach. We are aiming at extracting the fundamental properties from the large amount of data and revealing the influences of time effects on temporal networks from a new perspective.

The rest of the paper is organized as follows. Section II briefly introduces the method of factorial moments. In Section III we give a brief description of the data sets and present the corresponding results, especially the scaling properties of fluctuations for contacts and nodes in the empirical data sets. Conclusions are offered in the final section.

2. Method of factorial moments

Temporal network consists of a sequence of contacts (n_i, n_j, t), representing that node i and node j has a contact at time t. The number of contacts characterizes the frequency that individuals are connected with each other and the number of nodes describe the activeness that individuals are involved. In this paper, factorial moment approach will be used to study the fluctuations of active contacts and nodes in temporal networks as well as the corresponding scaling properties among these fluctuations.

Factorial moments are originally introduced in nuclear physics to study the multiplicity fluctuation of hadrons produced during the high energy collisions[9, 10]. The factorial moment approach provides a general and sensitive method to characterize whether the fluctuations in multiplicity distributions arise from statistical contributions or dynamical interactions. Here we will employ the method to study the multiplicity of contacts and nodes and uncover the dynamical nature of events in temporal networks.

Consider the time series of contacts (or nodes) $y(t)$, where t is the time that contacts happen and t ranges from 0 to T. We divide the whole time range T into M equal bins (the remainders are discarded). So the time interval in each bin is $\Delta t = T/M$. Within each bin window m ($m = 1, 2, ..., M$), denote the number of contacts (or nodes) as n_m. Of course, n_m fluctuates for different bin windows. To measure the fluctuations and correlations, the $q - th$ order factorial moment is introduced as,

$$\begin{aligned} f_q &= \frac{1}{M}\sum_{m=1}^{M} n_m(n_m-1)...(n_m-q+1) \\ &= \langle n_m(n_m-1)...(n_m-q+1)\rangle. \end{aligned} \tag{1}$$

In factorial moments, $f_1 = \langle n_m \rangle$ is the mean number of contacts (or nodes) under a certain bin size, averaged over all the bins m. Note that n_m must be greater than q ($n_m > q$) in order to contribute to f_q , and q is usually an integer. As M increases, Δt is decreased and the average multiplicity $\langle n_m \rangle$ in a bin decreases. This may lead to $n_m < q$ which is not allowed. Thus high q corresponds to higher n_m in the bin under consideration, i.e., large fluctuations from $\langle n_m \rangle$ [11].

Normalized factorial moments are more generally used,

$$F_q = \frac{f_q}{f_1^q}. \tag{2}$$

It was proved that F_q can filter out the statistical fluctuations [9, 10]. For uncorrelated statistical (Poissonian or Gaussian) distributions, $F_q = 1$ for all orders q. The method of factorial moments has been applied to analyze different complex systems, such as multiplicity of produced hadrons [12], human electroencephalogram and gait series in biology [13, 14], financial price series [15], critical fluctuations in Bak-Sneppen model [16], spectra analysis of complex networks [17], to name a few. Specially it exhibits that the system has self-similarity structures when F_q has a power-law dependence on the bin size M.

$$F_q \propto M^{\alpha_q}, \qquad \alpha_q > 0. \tag{3}$$

This phenomenon is referred to as the intermittency. Intermittency basically means random deviations from smooth or regular behavior and is expected in a variety of statistical systems at the phase transition point of the second-order type.

If F_q does not have an evident scaling behavior with M, the relationship between F_q and F_2 will be investigated,

$$F_q \propto F_2^{\beta_q}, \tag{4}$$

This is more widely valid for intermittent systems even though Eq. (3) does not hold [18, 19].

3. Results and discussions

In this paper the factorial moments analyses are performed to uncover the scaling properties of the fluctuations in temporal networks based on the following two empirical data sets.

Prostitution: The data set consists of sexual contacts between sex buyers and sellers from a Brazilian web forum [20]. The time resolution is 1 day and the whole time range is $T = 2232$ days.

Conference: The data set was collected at a 3-day conference from face-to-face interactions between conference participants. A contact is recorded every 20-second intervals if two individuals are within range of 1.5m [21]. The whole time range is $T = 212340$ seconds.

We now divide the whole time range T into M bins and count the number of contacts and nodes in each bin window. Calculate f_q and F_q according to Eq. (1) and (2), respectively. It is noticed that f_q is averaged over all bins (known as the horizontal average).

Figure 1 presents the time series of node and contact activities for *Prostitution* data. It is observed that the same data set has similar temporal patterns. We next go into the details of these time series. We plot in Fig. 1 the log-log graph of F_q as a function of M for contacts (open circles) and nodes (filled circles). With M ranging from about 3 to 60 bins, it means that the time interval Δt extends approximately from 30 to 750 days. The increase of bin number M means that the fluctuations of arbitrary sizes can appear in the system, and consequently leads to the growth of factorial moment F_q with M. In Fig. 1 we show that there is no power-law relationship between F_q and M of Eq. (3) over the whole range, but within some smaller ranges there is approximate linearity in each [22]. As stated in Ref. [15], for uncorrelated Poissonian or Gaussian distributions, $F_q = 1$ for all orders q; whereas for correlated distributions, F_q should increase with the growth of bin number M. Figure 1 suggests that the fluctuations of contacts and nodes in *Prostitution* data are not random Poisson distribution but have dynamical and

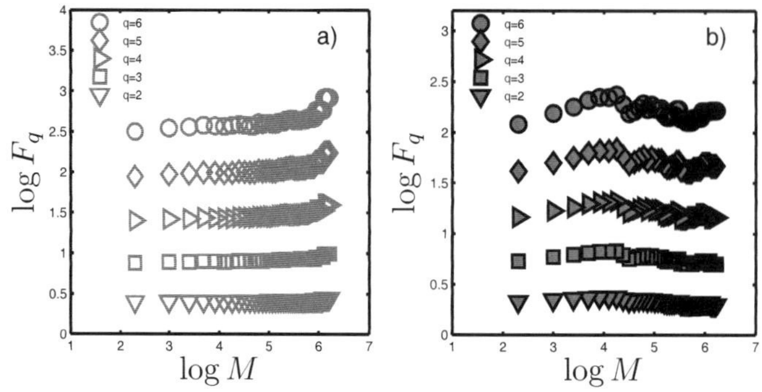

Figure 1. Log-log plot of factorial moments F_q as a function of bin size M for *Prostitution* data with q varying from 2 to 6. (a) The fluctuations of contacts (open circles). (b) The fluctuations of nodes (filled circles). The symbols are the same below.

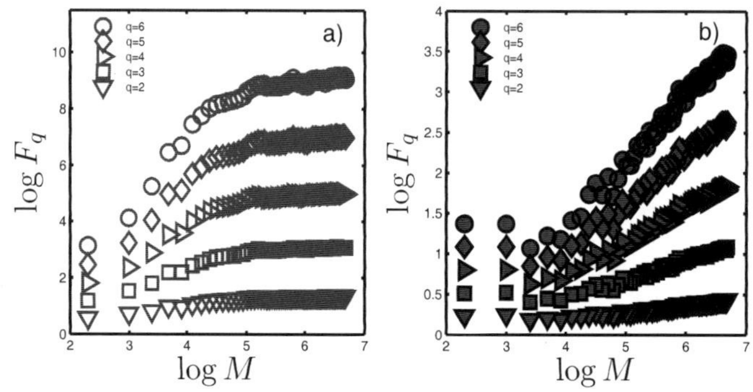

Figure 2. Log-log plot of F_q as a function of M for *Conference* data with the range of q from 2 to 6. (a) for contacts; (b) for nodes.

correlated behaviors inside. The same phenomena have also been observed in *Conference* data in Fig. 2.

Further investigations have been performed on F_q and F_2. We plot F_q as a function of F_2 on the log-log scale for *Prostitution* and *Conference* data sets in Fig. 3 and 4, respectively. The figures present remarkable linearity over the entire range. Thus we have a strict scaling relationship between F_q and F_2 for all q,

$$F_q \propto F_2^{\beta_q}. \tag{5}$$

The result also implies that β_q is independent of M, which suggests a common feature of scaling invariance in temporal networks.

Next we investigate the dependence of β_q on q. The plot of β_q as a function of $(q-1)$ is presented on a log-log scale in Fig. 5 for *Prostitution* data and in Fig. 6 for *Conference* data. There is also a strict linear relationship between β_q and $(q-1)$ for all q. Now one has

$$\beta_q \propto (q-1)^{\gamma}. \tag{6}$$

The linear fits are also plotted in the figures. In *Prostitution* data, γ are 1.44 for contacts and 0.629 for nodes. In *Conference* data γ are 1.211 and 1.367 for contacts and nodes, respectively.

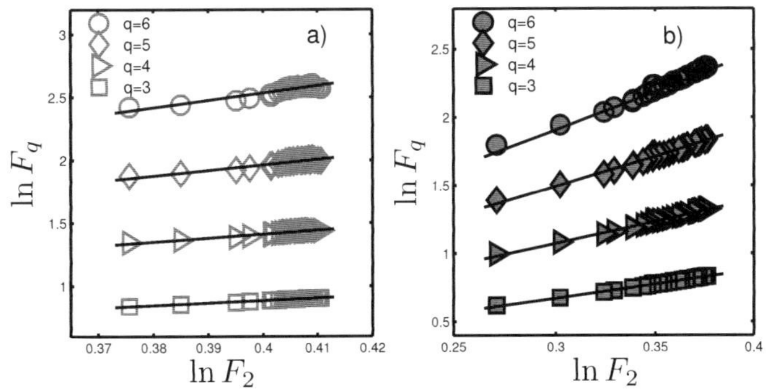

Figure 3. Log-log plot of F_q as a function of F_2 for *Prostitution* data. (a) for contacts; and (b) for nodes.

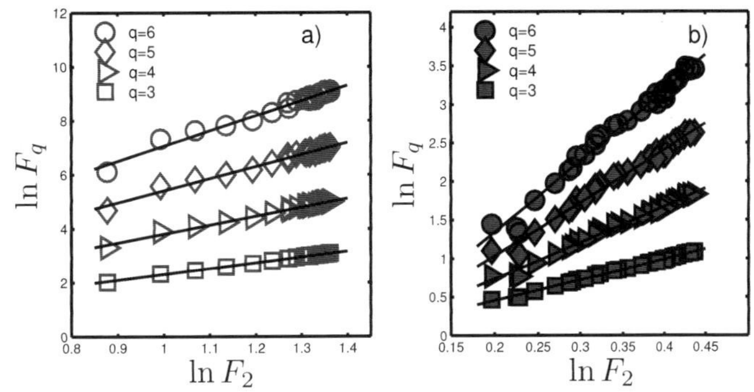

Figure 4. Log-log plot of F_q as a function of F_2 for *Conference* data. (a) for contacts; and (b) for nodes.

It is known that γ is 1.304 for systems undergoing second-order phase transition in Ginzburg-Landau (GL) theory and γ is independent of the details of the GL parameters [23, 24]. Except the exponent γ of node fluctuation for *Prostitution* data in Fig. 4(b), the other three exponents are all very close to 1.3. If we recall Fig. 2 for the measure of intermittency, we obtain that the more linear F_q and M are, the more likely the exponents γ are close to 1.3. The exponent γ indicates the strength of intermittent behavior in the dynamical fluctuations of temporal networks.

4. Conclusions

The method of factorial moments are performed to study the scaling properties of dynamical fluctuations in temporal networks based on empirical data sets of *Prostitution* and *Conference*. The phenomena of intermittency $F_q \sim M^{\alpha_q}$ over the whole range is not evident. But we have obtained a strict power-law relationship between F_q and F_2, $F_q \sim F_2^{\beta_q}$ for all orders q. The result implies that the fluctuations are not purely random, but have dynamical and correlated behaviors embedded in the system. We further find that β_q scales with q as $\beta_q \sim (q-1)^{\gamma}$. Some of the exponents γ are very close to that found in the system undergoing second phase-transition in GL theory. We suggest that the exponent γ describes the strength of intermittent behavior in the dynamical fluctuations of temporal networks.

Still, there are some issues to be addressed. First, what is the driving mechanism(s) behind

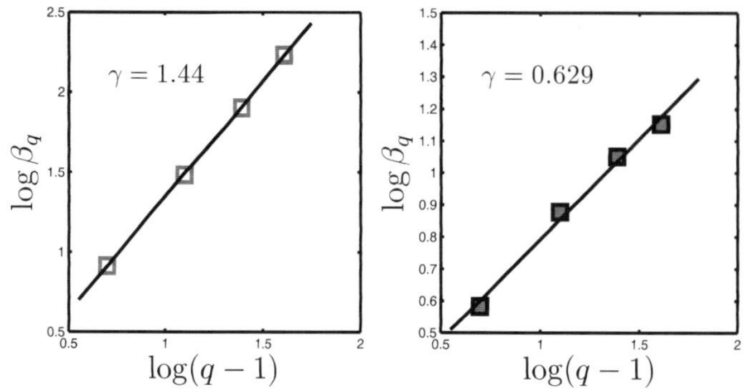

Figure 5. Scaling properties between β_q and $(q-1)$ for *Prostitution* data. The scaling exponents are about: (a) 1.341 for contacts; and (b) 1.104 for nodes.

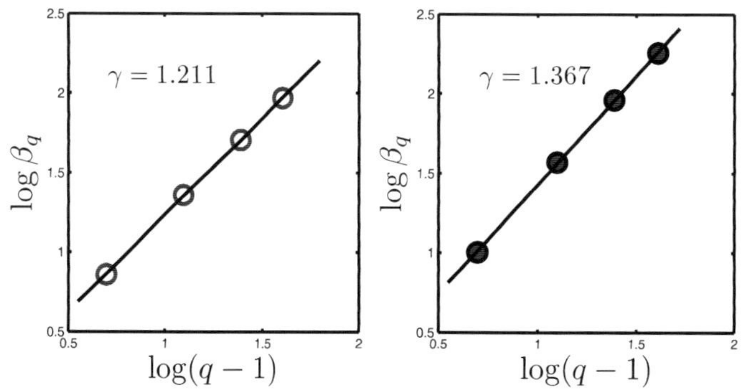

Figure 6. Scaling properties between β_q and $(q-1)$ for *Conference* data. The scaling exponents are about: (a) 0.992 for contacts; and (b) 1.345 for nodes.

these scaling properties of fluctuations in temporal networks? Second, why are some scaling exponents close to that of Ginzburg-Landau second-order phase transition? Are they belong to the same universal class? All these topics cannot be covered in this paper and will be discussed later.

The scaling invariances of fluctuations shed light on the temporal correlations of contact series and provide a new sight into understanding the influence of time in temporal networks.

Acknowledgments
We are grateful to Professor Petter Holme for providing the data. This work was supported by National Natural Science Foundation of China under No. 11605069, the programme of Introducing Talents of Discipline to Universities under No. B08033, and by the self-determined research funds of CCNU from the colleges' basic research and operation of MOE under Nos. CCNU2015A05046 and CCNU17QN0012.

References
[1] Holme P and Saramäki J 2012 *Physics Reports* **519** 97
[2] Trajanovski S, Scellato S and Leontiadis I 2012 *Phys. Rev. E* **85** 066105

IOP Conf. Series: Journal of Physics: Conf. Series **1113** (2018) 012004 doi:10.1088/1742-6596/1113/1/012004

[3] Krings G, Karsai M, Bernhardsson S, et al 2012 *EPJ Data Science* **1** 1
[4] Holme P 2013 *PLoS computational biology* **9** e1003142
[5] Barabási A-L 2005 *Nature* **435** 207
[6] Lambiotte R and Tabourier L 2013 *Eur. Phys. J. B* **86** 320
[7] Rocha L E C, Liljeros F and Holme P 2011 *PLos computational biology* **7** e1001109
[8] Miritello G, Moro E and Lara R 2011 *Phys. Rev. E* **83** 045102
[9] Bialas A and Peschanski R 1986 *Nucl. Phys. B* **273** 703
[10] Bialas A and Peschanski R 1988 *Phys. Lett. B* **207** 59
[11] Hwa R C 1998 *Phys. Rev. D* **57** 1831
[12] Yang C B and Cai X 1998 *Phys. Rev. C* **57** 2049
[13] Hwa R C and Ferree T C 2002 *Phys. Rev. E* **66** 021901
[14] Yang H J, Zhao F C, Zhuo Y Z, Wu X Z and Li Z X 2002 *Physica A* **312** 23
[15] Schoeffel L 2011 arXiv:1108.5946v1; available at SSRN: http://ssrn.com/abstract=1955253
[16] Xiao S S and Yang C B 2012 *Eur. Phys. J. B* **85** 154
[17] Zhao F C, Yang H J and Wang B H 2005 *Phys. Rev. E* **72** 046119
[18] Ochs W and Wosiek J 1988 *Phys. Lett. B* **214** 617
[19] Pan J and Hwa R C 1992 *Phys. Rev. D* **46** 4890
[20] Rocha L E C, Liljeros F and Holme P 2010 *Proc. Natl. Acad. Sci.* **107** 5706
[21] Isella L, Stehlé J, Barrat A, Cattu C, Pinton J F and van den Broeck W 2011 *J. Theor. Biol.* **271** 166
[22] L. Chi and C. Yang, Proceedings of Nonlinear Theory and its Applications, Hongkong, China (2015) 137.
[23] Hwa R C and Nazariv M T 1992 *Phys. Rev. Lett.* **69** 741
[24] Hwa R C and Pan J 1992 *Phys. Lett. B* **297** 35

IOP Conf. Series: Journal of Physics: Conf. Series **1113** (2018) 012005 doi:10.1088/1742-6596/1113/1/012005

Relationship between the degree exponent and its associated timeseries' hurst exponent of a static scale-free network — Under the viewpoint of mutual representation between timeseries and network graph

Jie Liu, Hongling Liu, Jian Jiang, Bengong Zhang*

Research Centre of nonlinear Science, Wuhan Textile University, Wuhan, Hubei, 430073, P. R. China.

E-mail: liujie_hch@163.com, benyan1219@126.com (corresponding author)

Abstract. During the past decade, bridging the gap between time series and complex networks, which aims to investigate the usage of network features to depict geometry of different type of timeseries becomes a hot topic. In this brief paper, the internal relationship between a typical static scale-free networks and its associated nonlinear timeseries (under the viewpoint of duality analysis of networks and time series) was investigated. A ubiquitous process to discuss such a problem are introduced. It is found that, the scale-free degree exponent has linear relationship with its associated timeseries' hurst exponent. Research in this direction will provide new methods and powerful tools for analysis of the causal relationships between complex structures and temporal behaviors of a complex dynamical system. Also, it is useful for further investigation on structure and function analysis of complex networks via classical timeseries analysis methods.

1. Introduction

In recent several years, several types of new mapping methods had been made to bridge the gap between time series and complex networks, which aims to study the usage of network features to depict geometric features belongs to different types of timeseries. Some important studies have shown that, different types of timeseries are corresponding to different types of topological structures. What relationship between these topology characters (ie.,the network characteristic quantities) and the original time series, is still unclear. How to apply the network analysis methods to depict the characteristics of the original timeseries' statistics (vice versa the inverse process of such a research), therefore, become extremely urgent nowadays. After some recent published literatures showed that, different types of time series is corresponding to different types of topological structure, this theme quickly becomes a hot point in nonlinear science and network science research field [1]-[7].

What relationship between these topology characters (ie.,the network characteristic quantities) and the original time series, is still unclear. How to apply the network analysis method to depict the characteristics of the original time series' statistics (vice versa the inverse process of such a research) therefore become extremely urgent nowadays.

Under the viewpoint of duality analysis of networks and timeseries, we will addressed on the internal relationship between a typical static scale-free network and its associated nonlinear timeseries in this brief paper. We will discuss on the relationship between scale-free degree exponent and its associated timeseries' hurst exponent.

The rest of this paper is arranged as follows: in section 2, some backgrounds will recalled firstly, including the static model with adjustable degree exponents, the concept of hurst exponent, and constructing associated timeseries methods by using recurrence plot matrix and classical multi-dimensional scale analysis method. In section 3, we will give main results by using the ubiquitous process for discussing such problem proposed in section 2. In section 4, brief conclusions are finally concluded. Research in this direction will provide new methods and powerful tools, for analysis of the causal relationships between complex structures and temporal behaviors of a complex system, and as well as for structure and function analysis of a complex networks.

2. Static scale-free network with scalable scale exponent, hurst exponent and duality transfer method based on recurrence plot matrix

In this section, we will firstly introduce the static scale-free network with scalable scale index [8], Hurst index [12], and duality transformation method [1]-[7], which will be used in following discussion.

2.1. Recall the related static scale-free network model(SM)

In Ref. [8], a static model (registered as SM in the following context) was firstly introduced as an non-increased network algorithm to generate static network (i.e. not growing) with typical scale-free index (with any desired degree exponent γ larger than or equal to 2). Firstly, let's recall the model defined as follows [9]:

The SM is started from N disconnected vertices indexed by an integer number $i \in N$, taking the values $i = 1, \dots N$. A normalized probability p_i is assigned to each vertex given as function of the index i by

$$p_i = \frac{i^{-\alpha}}{\sum_{j=1}^{N} j^{-\alpha}}, \tag{1}$$

where α is a real number in the range $\alpha \in [0, 1]$. The following rules are obeyed by the algorithm:

- The SM network is constructed iterating two different vertices i and j are randomly selected from the set of N vertices, with probability p_i and p_j, respectively. If there exists an edge between these two vertices, they are discarded and a new pair is randomly drawn. Otherwise, an edge is created between vertices i and j.

- Repeated such a process until $E = mN$ edges are created in the network, accounting for a fixed average degree $\langle k \rangle = 2E/N = 2m$.

In this algorithm, a static network was generated with no self-connections (a vertex joined to itself) not multiple connections (two vertices connected by more than one edge). By means of a simple mean-field argument mentioned in [8] and [9], one can analyze the corresponding degree distribution of this **SM** easily.

Since edges are connected to vertices with a probability given by the factor p_i, the probability of any edge belongs to the vertex i whose degree equals to k_i) is:

$$\frac{k_i}{\sum_j k_j} \sim p_i. \tag{2}$$

For the case of $N \to \infty$, approximating sums by integrals, for $0 < \alpha < 1$, one can see (3),

$$\sum_{j=1}^{N} j^{-\alpha} \sim \int_{1}^{\infty} j^{-\alpha} dj \sim \frac{N^{1-\alpha}}{1-\alpha}. \tag{3}$$

Since $\sum_j k_j = \langle k \rangle N$, from Eq. (2), one can see the following:

$$k_i \sim p_i \sum_j k_j \sim 2m(1-\alpha)\left(\frac{i}{N}\right)^{-\alpha}. \tag{4}$$

By using general arguments from network theory [10], one can obtain that, the degree distribution relationship to characterize the SM network have the form,

$$P(k) \sim k^{-\gamma}$$

with a associated scale free degree exponent as a function of parameter α:

$$\gamma = 1 + \frac{1}{\alpha}. \tag{5}$$

Obviously, by tuning the parameter α in the interval $[0, 1]$, the scale index of the generated SM network must behaved as a typical scale free network with a associated degree exponent, in the board range $\gamma \in [2, \infty)$.

Since the above mentioned algorithm generates a scale free networks with built-in degree correlations mechanism, the maximum degree corresponding to the index $i = 1$ can be calculated as:

$$k_{i=1} \sim 2m(1-\alpha)N^{\alpha}. \tag{6}$$

which implies that the cut-off (or maximum expected degree) $k_c(N)$ in the SM network [11] scales with the network size as $k_c(N) \sim N^{\alpha}$. So, in the case of absence of multiple and self-connections, in order to have no correlations, a scale-free networks with size N must have a cut-off scaling at most as $k_s(N) \sim N^{1/2}$ (the so-called structural cut-off) [11].

Therefore, the SM will yield a correlated network for given parameter values $\alpha > 1/2$, i.e., for degree exponents in the interval $2 < \gamma < 3$, which correspond to those values empirically observed in real networks with scale free characters.

In the following Sections, we will discuss the relationship between the scale index and its associated timeseries's hurst index of a correlated static scale-free network model (with degree exponents in the interval $2 < \gamma < 3$).

2.2. Recall the calculation of the Hurst exponent of timeseries

'Hurst exponent' or 'Hurst coefficient', derives from Harold Edwin Hurst (1880 – 1978), who was the lead researcher in this direction. The so called 'Hurst exponent' is used as a important measure of long-term memory of linear or nonlinear time series [12] [13]. It relates to the autocorrelations of the time series, and the rate at which these decrease as the lag between pairs of values increases. Studies involving the Hurst exponent were originally developed in hydrology for the practical matter of determining optimum dam sizing for the Nile river's volatile rain and drought conditions that had been observed over a long period of time.

The hurst exponent, H, is defined in terms of the asymptotic behaviour of the re-scaled range as a function of the time span of a time series as follows [12]:

$$E\left[\frac{R(n)}{S(n)}\right] = Cn^{H} \qquad (n \to \infty),$$

where $R(n)$ is the range of the first n values, and $S(n)$ is their standard deviation, $E[x]$, is the expected value. Parameter n is the time span of the observation (number of data points in a time series), and C is a constant.

The values of the Hurst exponent vary between 0 and 1, with higher values indicating a smoother trend, less volatility, and less roughness.

A number of estimators of long-range dependence of time series have been proposed in published literatures. The oldest and best well known is the so-called re-scaled range (R/S) analysis popularized by Mandelbrot and Wallis' milestone work, which was constructed based on previous hydrological findings of Hurst. Alternatives include DFA, periodogram regression, aggregated variances, local Whittle's estimator, wavelet analysis, both in the time domain and frequency domain. One can consult methods used in different fields from published literatures, i.e., [12]-[18], to name just a few.

Table 1. Significance of different value range of Hurst index

$H > 0.5$	indicating a smoother trend
$H = 0.5$	indicating a less volatility
$0 \leq H < 0.5$	indicating a less roughness

2.3. A typical transfer method to reproduce the associated timeseries of network graph

In refs [2],[3],[4], a method was widely used to reproduce distance matrices and original time series from recurrence plots was proposed by Hirata et al. Their procedure can fulfill the task to convert a recurrence plot matrix to a weighted network graph, and then calculate a distance matrix named by 'reproduced distance matrix' between each pair of nodes on this weighted graph. Their method can reproduce the topological shape of original time series fairly well. Such an effective algorithm will be used here, and one can find algorithm details in references [2],[3],[4], and [22].

Here comes the main step of this algorithm used in our numerical analysis parts:

- I) Prepare a recurrence plots(RP) of the considered nonlinear timeseries [19],[20],[21]: Eckmann et al., introduced recurrence plots, which provide a way to visualize the periodic nature of a trajectory through a phase space. It can be used to transfer a timeseries to network graph [22]. Let $x(i)$ be a state at time index $i(1 \leq i \leq n)$. Denote by R a function of two states that tells whether the two states are close or not. If they are close to each other, then R returns 1. Otherwise, it returns 0. Then a RP can be defined as $\{(i,j) : R(x_i, x_j) = 1, i,j = 1,2,\cdots,n\}$ (For more details on RP analysis method, one can consult references [19],[20],[21], to name just a few).
- II) Construct a graph from the RP: In such a graph, each node corresponds to a time index, and an edge exists between nodes i and j if the states x_i, x_j corresponding to the time indices i,j are close, namely, $R(x_i, x_j) = 1$.
- III) For each edge, assign a weight: Define G_i by $\{j : R(x_i, x_j) = 1\}$. G_i is the set of time indices to which i is close. For each existing edge between i and j, define the weight w by

$$w(i,j) = 1 - \frac{|G_i \bigcap G_j|}{|G_i \bigcup G_j|}, \tag{7}$$

  where $|A|$ shows the number of elements in set A, $\bigcap$ shows the intersection of two sets, and $\bigcup$ shows the union of two sets.
- IV) Calculate the shortest distance between each pair of nodes on the graph weighted by $w(i,j)$. (One can use the fast Dijkstra method for calculating the shortest distance.)

- V) Use classical multidimensional scaling analysis(**CMDS**) technique, which can arrange a set of points so that these points preserve the given distance matrix, to reproduce the associated timeseries. (The **CMDS** algorithm can be fulfilled by function cmdscale.m by Matlab software suite.)

In this algorithm, the classical multidimensional scaling method was used. In the reproduction process, the component with the largest eigenvalue are used. This algorithm can be seen as an extended version of classical Isomap method, which was widely used for dimension reduction except the definition of edge weights by Equation (7) is constructed in a new way. This method works fairly well with the condition that the related graph is connected. In the following sections, the above mentioned algorithm from step II to the end will be used for further numerical discussion.

3. Main results

In this section, we begin to discuss the topic we focus on: analysis relationship between the scale free degree exponent and its associated timeseries's Hurst exponent of a static network model. Since there are not a one-one map published for analyzing the problem theoretically yet. We will discuss the problem numerically. The basic procedure is listed as follows:

- Step I) Create a group of static scale free networks with the same scale N and average degree m;
- Step II) Create the associated timeseries of step I by using the aforementioned **CMDS** method;
- Step III) Calculating the scale-free degree exponent and its associated timeseries's hurst exponent and try to find hidden relationship between these two characters.

Results in details will be given in the followed subsections.

3.1. Numerical results and discussion

In Figure 1, we give the numerical results of relationship between scale index γ and Hurst index H with $N = 600$. It can be seen that, the linear mode relationship does exist there. The related information of this fitting process with N=600, is listed in Table 2. One can see that the goodness parameter R^2 of this fit is very close to 1, which means the curve fitting is linear mode.

Table 2. The related information of this fitting process $N = 600$, $m = 4$ for Figure 1(Coefficients with 95% confidence bounds.)

Linear model Poly1:	$f(x) = p_1x + p_2$
Coefficients :	$p_1 = 0.08006(0.0148, 0.1453), p_2 = 0.2346(0.07579, 0.3934)$
Goodness of fit:	SSE: 0.001541, R^2: 0.7436, Adjusted R^2: 0.6795, RMSE: 0.01963

In Figure 2, we give the numerical results of relationship between scale index γ and Hurst index H with $N = 900$. It can also be seen that, the linear mode relationship exists there, too. The related information of this fitting process $N = 900$, $m = 4$, was listed in Table 3. One can also see that the goodness parameter R^2 of this fit is very close to 1, which means the curve fitting is linear mode.

Similarly, in Figure 3, the numerical results of relationship between scale index γ and Hurst index H with $N = 1200$ is given. It can be seen that, the linear mode relationship exists there, too. The related information of this fitting process N=1200, m=4, are listed in Table 4. One

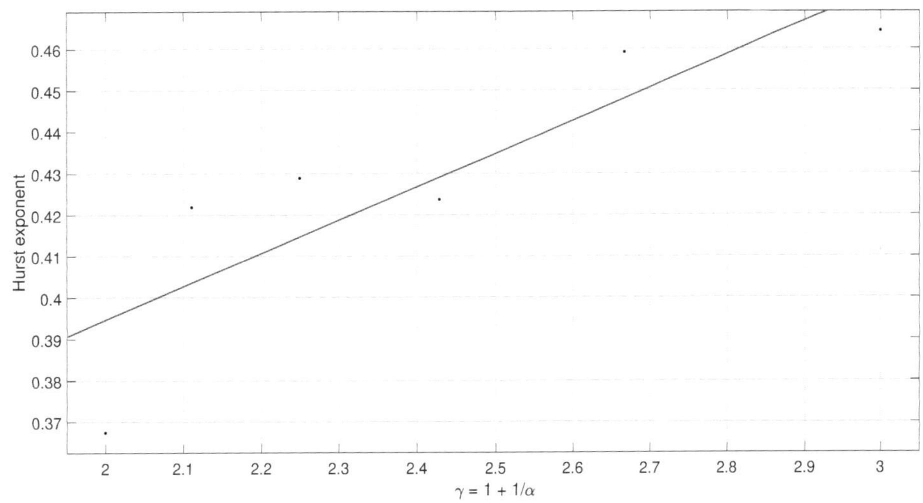

Figure 1. Relationship between scale index γ and Hurst index H, where $N = 600$. (mean of 20 run numerical experiments.)

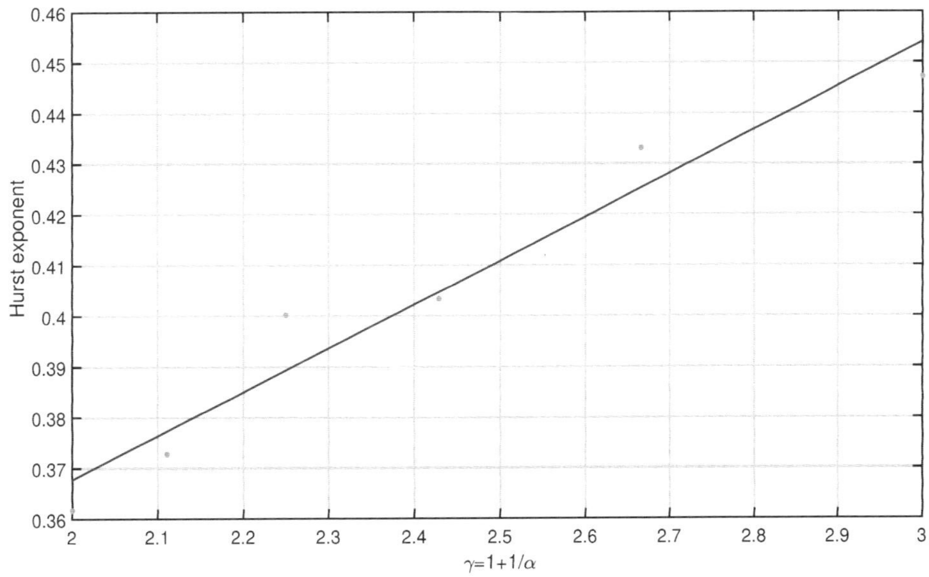

Figure 2. Relationship between scale index γ and Hurst index H, where $N = 900$. (mean of 20 run numerical experiments.)

can also see that the goodness parameter R^2 of this fit is very close to 1, which means the curve fitting is linear mode.

Finally, in Figure 4, we gave the numerical results of relationship between scale index γ and Hurst index H with $N = 1500$. It can be seen that, the linear relationship exists there, too. The related information of this fitting process N=1500, m=4, are listed in Table 5. One can also

Table 3. The related information of this fitting process N=900, m=4 for Figure 2(Coefficients with 95% confidence bounds.)

Linear model Poly1:	$f(x) = p_1x + p_2$
Coefficients :	$p_1 = 0.08625(0.05807, 0.1144), p_2 = 0.1952(0.1266, 0.2638)$
Goodness of fit:	SSE: 0.0002874, R^2: 0.9475, Adjusted R^2: 0.9344, RMSE: 0.008477

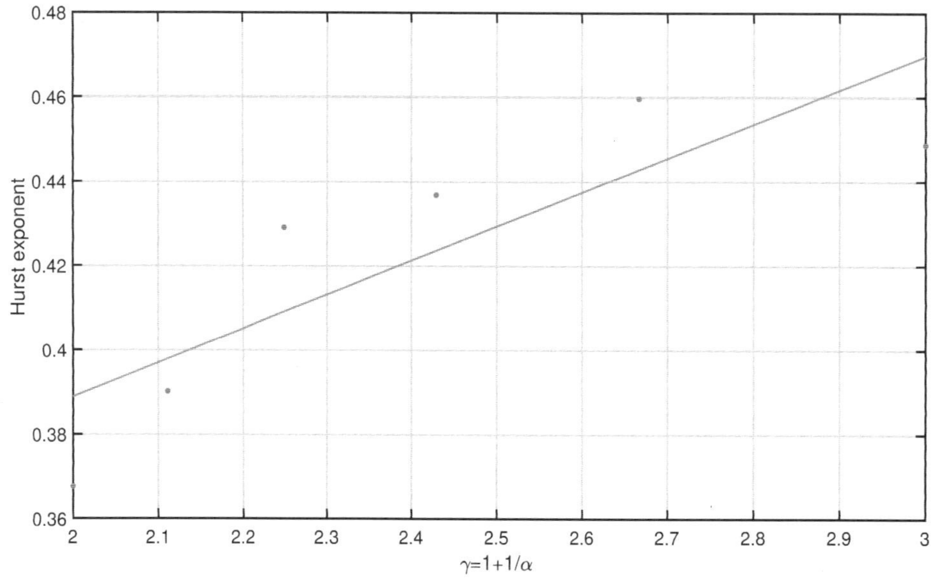

Figure 3. Relationship between scale index γ and Hurst index H, where $N = 1200$. (mean of 20 run numerical experiments.)

Table 4. The related information of this fitting process N=1200, m=4 for Figure 3,(Coefficients with 95% confidence bounds.)

Linear model Poly1:	$f(x) = p_1x + p_2$
Coefficients:	$p_1 = 0.08088(0.01024, 0.1515), p_2 = 0.2271(0.05526, 0.399)$
Goodness of fit:	SSE: 0.001805, R^2: 0.7164, Adjusted R^2: 0.6455, RMSE: 0.02125

see that the goodness parameter R^2 of this fit is very close to 1, which means the curve fitting is linear mode. We also gave the residuals plot (fitting error bar plot), from which, one can see the fitting behaves fairly well.

Table 5. The related information of this fitting process N=1500, m=4 for Figure 4,(Coefficients with 95% confidence bounds.)

Linear model Poly1:	$f(x) = p_1x + p_2$
Coefficients:	$p_1 = 0.1043(0.07738, 0.1313), p_2 = 0.1637(0.09818, 0.2293)$
Goodness of fit:	SSE: 0.0002627, R^2: 0.9665, Adjusted R^2: 0.9582, RMSE: 0.008104

From the above mentioned numerical results, one can see that, with the increasing of network graph scale (i.e., the length of related associated timeseries) the linear mode relationship becomes

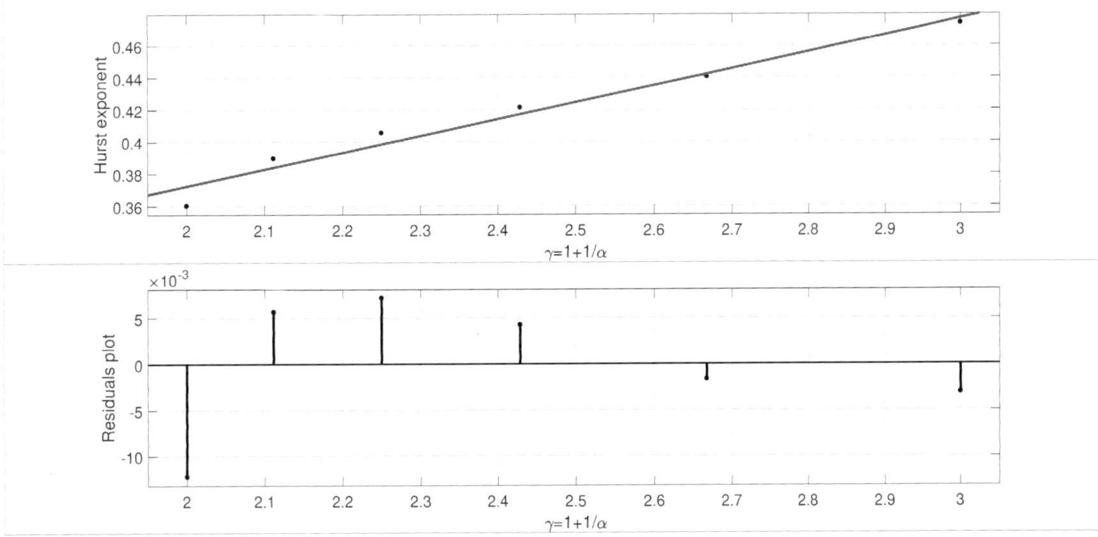

Figure 4. Relationship between scale index γ and Hurst index H, where $N = 1500$. (mean of 20 run numerical experiments.)

more and more clearly (long term tendency). Limited to the processing power of our personal computers, this paper fails to provide large-scale numerical results. But we infer that this linear mode correlation phenomenon might exist there.

Also, we found that, the Hurst exponents are all small than 0.5 under different network graph scale, which means that, the associated timeseries created from static scale-free network with adjustable degree exponent $\gamma \in [2, 3]$ are with less roughness. This is coincide with the real character of real nonlinear timeseries which induced the related networks [1], [2].

3.2. Compare the fit tendency

In this section, we further give the fit tendency analysis. In the following figure 5, the relationship of network graph scale and the R-Square parameter R^2 is listed. We fitted the tendency by a linear-like mode function $f(x) = a(sin(x - \pi)) + b((x - 10)^2) + c$, where the related coefficients with 95% confidence bounds, are listed as follows, respectively.

$$\begin{aligned} a &= -0.05591 \quad (-2.595, 2.483) \\ b &= 8.051e-08 \quad (-2.467e-06, 2.628e-06) \\ c &= 0.7668 \quad (-2.485, 4.018). \end{aligned}$$

One can see that, the R^2 parameter is roughly increasing with network graph scale parameter N (the oscillation might be caused by numerical error, it might be smoothed under more times numerical run average), which means that, the larger network graph scale is, the more definitely of existing linear mode relationship between scale-free degree exponents and hurst exponents of its associated timeseries. So, one might deduce that, there do exist some 'one-one map' between network graph and its associated timeseries, and vice vera.

4. Conclusion

Under the viewpoint of duality analysis of networks and time series, the internal relationship between a typical static scale free networks and its associated nonlinear timeseries are investigated in this brief paper. It is found that, the scale-free degree exponent has linear mode

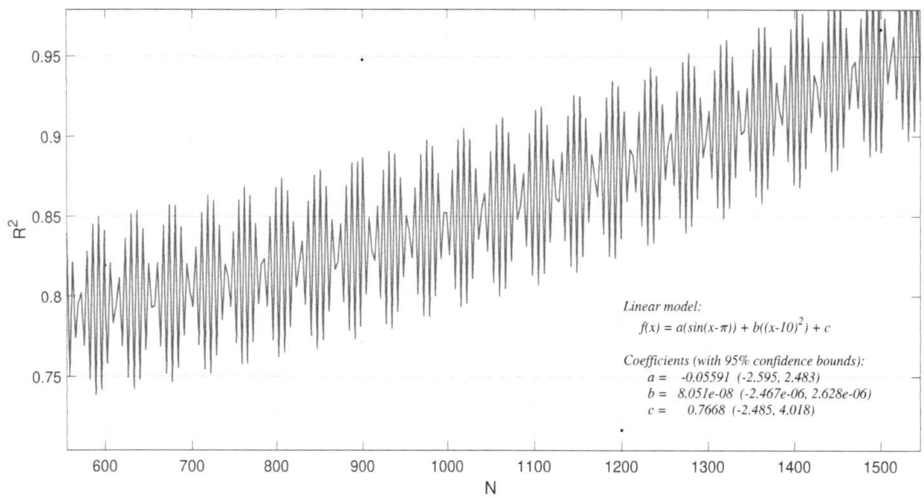

Figure 5. Relationship between N and R^2, where $N = 600, 900, 1200, 1500$.

relationship with its associated timeseries' Hurst exponent. The larger network graph scale is, the more definitely of existing linear mode relationship between scale free degree exponent and Hurst exponent of its associated timeseries. So, one can deduce that, there do exist some 'one-one map' between network graph and its associated timeseries, and vice vera. Research in this direction will provide new methods and potential tools, for analysis of the causal relationship between complex structures and temporal behaviors of a complex dynamical system, and as well as for structure and function analysis of a complex network graph. In the future, we will further investigate the relationship between other network characters and the statistical character of the associated timeseries. It might reveal some deep relationship for duality analysis of nonlinear timeseries and complex network graph.

Acknowledgement
This work is supported by National Natural Science Foundation (NNSF) of China under Grant No.61573011.

References
[1] Shirazi A H. Jafari G R. Davoudi J. Peineke J. Tabar M. and Muhammad S. 2009 Mapping stochastic processes onto complex networks *J of Statis Mech* **7** P07046.
[2] Hirata Y. Horai S. and Aihara K. 2008 Reproduction of distance matrices and original time series from recurrence plots and their applications *Eur. Phys. J. Special Topics* **164** pp 13-22.
[3] Shimada Y. Ikeguchi T. and Shigahara T. 2012 From networks to time series *Phys Rev Lett* **109** 158701.
[4] Haraguchi Y. Shimada Y. Ikeguchi T. and Aihara K. 2009 Transformation from complex networks to time series using classical multidimensional scaling. *Proc of the 19th Int Conf on Artif Networks* Springer-Verlag.
[5] Strozzi F. Zaldivar J. M. Pioljansek K. Bono F. and Gutierrez E. 2009 From complex networks to time series analysis and viceversa: Application to metabolic networks. *EUR-Scientific and Technical Research series* (Luxembourg) pp 1-56.
[6] Campanharo A S L O. Sirer M I. Malmgren R D. Ramos F M. and Amaral L A N. 2011 Duality between Time Series and Networks. *PLoS ONE* **6** (8) e23378.
[7] Zhao Y. Weng T F. and Ye S. 2014 Geometrical invariability of transformation between a time series and a complex network. *Physical Review E* **90**(1) 012804.
[8] Goh K I. Kahng B. and Kim D. 2001 Universal behavior of load distribution in scale-free networks. *Physical Review Letters* **87**(27 Pt 1) 278701.

[9] Catanzaro M. and Pastor-Satorras R. 2005 Analytic solution of a static scale-free network model. *The European Physical Journal B - Condensed Matter and Complex Systems* 2005, **44**(2) pp 241-248.
[10] Dorogovtsev S N. and Mendes J. F. F. 2003 *Evolution of networks: From biological nets to the Internet and WWW* (Oxford: Oxford University Press, Inc)
[11] Dorogovtsev S N. and Mendes J F F. 2003 *Evolution of Networks: From Biological Nets to the Internet and WWW (Physics)* (Oxford: Oxford University Press, Inc)
[12] Hurst H E. 1951 Long term storage capacity of reservoirs. *Transactions of the American Society of Civil Engineers* **116**(12) pp 776-808.
[13] Lloyd E H. Hurst H E, Black R P. et al. 1966 Long-Term Storage: An Experimental Study. *Journal of the Royal Statistical Society* **129**(4) pp 591.
[14] Mandelbrot B B. and Wallis J R. 1968 Noah, Joseph, and Operational Hydrology. *Water Resources Research* **4**(5) pp 909-918.
[15] Mandelbrot B B. Wallis J R. 1969 Robustness of the rescaled range R/S in the measurement of noncyclic long run statistical dependence. *Water Resources Research* **5**(5) pp 967-988.
[16] Fowler A D. 1991 Fractals : by J. Feders, Plenum Press, New York, 183. *Computers and Geosciences* **17**(7) pp 1065-1066.
[17] Ray W D. 1994 Statistics for Long-Memory Processes. by J. Beran *Statistics for long-memory processes* (Chapman and Hall) pp 105-106.
[18] Doukhan P. 2003 *Theory and applications of long-range dependence* (Birkhuser)
[19] Eckmann J. P. Kamphorst S. O. and Ruelle D. 1987 Recurrence Plots of Dynamical Systems *Europhysics Letters* **5** (9) pp 973977.
[20] Marwan N. Romano M. C. Thiel M. and Kurths J. 2007 Recurrence Plots for the Analysis of Complex Systems *Physics Reports* 438 (5-6)pp 237.
[21] Marwan N. 2008 A historical review of recurrence plots *European Physical Journal ST* **164** (1) pp 312.
[22] Liu J. Shi S-T. and Zhao J-C. 2013 Comparison study of typical algorithms for reconstructing time series from the recurrence plot of dynamical systems. *Chinese Physics B* **22**(1) pp 131-137.
[23] Liu J. Liu H-L. Huang Z-J and Tang Q. 2015 Differ multivariate timeseries from each other based on a simple multiplex visibility graphs technique. *International Conference on Intelligent Control and Information Processing* IEEE pp 289-295.
[24] Liu J. and Li Q-Q. 2017 Planar Visibility Graph Network Algorithm For Two Dimensional Time series, Proceeding of 2017 CCDC, Chongqing, China. 2017. May. 28th - 2017. May. 30th.
[25] Liu J. 2017 Research on dynamics and structures' mutual representation between nonlinear time series and complex network graphs. *Proceedings of WANCSA 2017* (Normandie University, Le Havre, France. July. 4th -5th) pp 1-2.

The Spatial SIS model of epidemic spreading on networks

Zi Hui*

School of Electrical and Electronic Engineering, Wuhan Polytechnic University, Wuhan 430032, China

E-mail: huizi0117@qq.com

Abstract. In this paper, we define a Spatial Susceptible-Infectious-Susceptible model for discussing how disease spreads on spatial driven network. The spreading rate is affected by spatial distance of links and infective medium. Numerical studies exhibit that infective medium is bad for disease spreading. With the increasing of β, spreading behavior needs longer time to reach to the steady state, but reaches to smaller steady infected ratio. In addition, also by numerical study, when spreading rate is small, epidemic behavior is affected by some special nodes.

Keywords: S-SIS model, spatial driven network, epidemic spreading

1. Introduction

The behavior of epidemic spreading has been described by several models[1-3]. In 1975, Bailey introduced a model, which was known as the susceptible-infectious-susceptible (SIS) model[1]. In this model, individuals could be divided into two classes: susceptible (S) individuals who don't have the disease yet but can catch it if they contact to someone who does, and infective (I) individuals who have the disease and can pass it on. In SIS model, infective individuals recover to susceptible individuals automatically after a few steps.

The SIS model cannot be solved exactly on a network, but Pastor-Satorras and Vespignani used mean-field theory to solve SIS model on scale-free network[4,5]. In these works, each of the links has equal effective spreading rate under mean-field approximation. The epidemic behavior is completely determined by degree distribution. The SIS model can also be extended to describe the real epidemic spreading behavior to predict the infection density of disease propagation and to reduce the duration of epidemic outbreak[6,7].

However, many real systems are embedded in space[8-11]. In such real systems, spatial distances between individuals play important role not only in structure but also in dynamical process. Several spatial network models have been proposed for studying the spatial characters and the spatial dependence of dynamical process[12-15]. Scientists have revealed two important phenomena. The first one is on average the majority of our social relationships are in our neighborhood[15]. The second one is the disease has higher probability to spread on short spatial distance than on long spatial distance. Therefore, spatial properties are of great significance to study the epidemic spreading.

In this paper, we study the epidemic spreading process on growing spatial driven network[13]. We focus on how network spatial structure and spreading rate influence spreading behavior. The spreading rate is not a constant, but it is inversely proportional to the power function of spatial distance for each link. Monte Carlo method is used in the simulation of dynamics.

2. Spatial Susceptible-Infectious-Susceptible Model

In SIS model[1], each node in the network represents an individual, and each link represents a connection. The disease spreads from one node to another through a link. These nodes have two states: susceptible and infected. At each time step, each susceptible node is infected with rate ν if it links with one or more infected nodes. At the same time, infected nodes are cured and become susceptible again with rate δ. Thus, the effective spreading rate $\lambda = \nu/\delta$. Simplified, we always set $\delta = 1$. According to the mean-field approximation, in the traditional SIS model, each link has the same spreading rate λ. But in real systems, the disease always spreads through human contact, air and other infective media. Most of these factors are affected by spatial distance. Thus, we give the definition of Spatial Susceptible-Infectious-Susceptible (S-SIS) Model:

We assume that λ_0 denotes the infective rate of the disease itself. The cure rate $\delta = 1$ for simplified. Inspired from previous works[16-18], we define the effective spreading rate λ_{sj} from infective node j to its susceptible neighbor s is inversely proportional to their spatial distance as follows :

$$\lambda_{sj} = \lambda_0\left(\frac{min_{k\in\Omega_j} d_{kj}}{d_{sj}}\right)^{\beta}, \tag{1}$$

where Ω_j is the set of all susceptible neighbors of node j. The distance between infected node and its nearest susceptible neighbor is used to rescale λ_{sj} $(0 < \lambda_{sj} \leq \lambda_0)$. Thus, the spreading rate from infective node to its nearest susceptible neighbor is λ_0. For other susceptible neighbors, the spreading rate is inversely proportional to their spatial distance with power β. We assume β is the parameter of infective medium. For example, malaria spreads by mosquitoes. Many empirical studies and theoretical analysis on the pattern of animal foraging and migration have uncovered that the mobility travel step size for animal follows a power-law[19-21]. What's more, the step size of human traveling follows a power-law as well. For smaller β, the medium is easier to spread disease; for larger β it is more difficult.

In S-SIS model, the time-discrete dynamics is defined by synchronously updating the states of all individuals with the following rules: If node s is infected at time $t-1$, it is susceptible at time t. Otherwise, if node s is susceptible at time $t-1$,and it connects to one or more infected nodes at the same time, then node s is infected at time t with probability λ_s. The value of λ_s depends on all infective neighbors and can be written as

$$\lambda_s = 1 - \prod_{k\in\varphi_s}(1-\lambda_{sk}) \tag{2}$$

φ_s is the set of all infective neighbors of node s. In order to discuss how network spatial structure and infective medium influence the epidemic behavior, we set $\lambda_0 = 1$.

3. Numerical Results

3.1. Spatial driven network

The individuals are connected by spatial driven networks[13]. The network is fixed on the plan, each of the nodes has a two-dimensional coordinate (x, y) $(0 \leq x < 1, 0 \leq y < 1)$. The networks are constructed by adding nodes and links in the following way: In initial state $(t = m_0)$, we generate $m_0 + 1$ all-to-all connected nodes. At every time step, a new node n is randomly located in (x_n, y_n) on the plan. Node n connects with m previous nodes, which are selected by the probability $\pi_i = \frac{d_{ni}^{-\alpha}}{\sum_j {d_{nj}}^{-\alpha}}$ $(0 \leq \alpha)$. d_{ni} is the spatial distance between new node n and previous node i. Close the iteration loop by generating the next new node and so forth, until the network size reaches the desired value $N = 10000$. The degree distribution grows exponentially and is determined by m as follows:

$$p(k) = \frac{1}{m+1}(\frac{m}{m+1})^{k-m}, (k \geq m). \tag{3}$$

In this paper, we set $m_0 = 2$ and $m = 2$, which means that spatial networks with different α have the same degree distribution. The spatial distance distribution follows $p(d) \sim d^{-(\alpha-1)}$.

3.2. S-SIS model on spatial driven networks

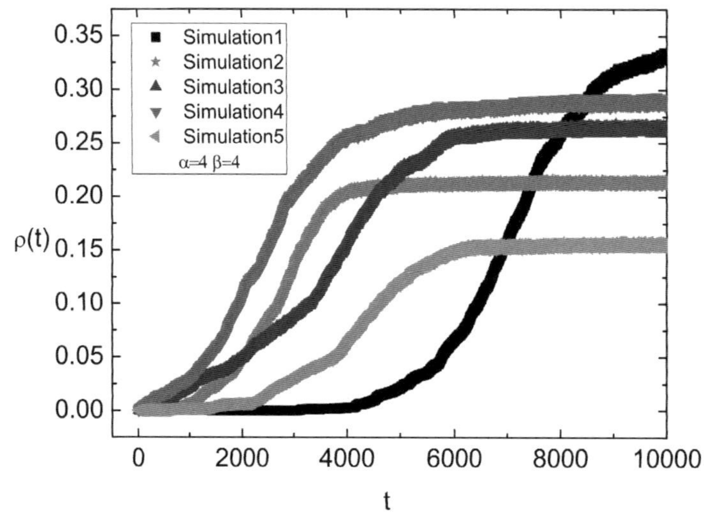

Figure 1. The infected nodes ratio $\rho(t)$ as a function of time t with $\alpha = 4$ and $\beta = 4$. Different symbols represent different simulations.

The disease spreads in spatial driven networks following the S-SIS model. For each β in S-SIS model, we do 20 simulations on the same spatial driven network. In every simulation, we start by selecting one node randomly and assume it is infected. Figure 1 shows that, when $\alpha = 4$ and $\beta = 4$, the steady infected nodes ratio ρ_s is different for different simulations. ρ_s is the average of infected nodes ratio $\rho(t)$ from $t = 9000$ to $t = 10000$. The only difference for these 5 simulations in figure 1 is the first infected node. In this case, the initial state of infected node affects the spreading behavior. But in traditional SIS models, ρ_s has nothing to do with the first infected node.

We also discuss this phenomenon with $\alpha = 1, 2, 3, 4$ and $\beta = 1, 2, 3, 4$ in figure 2. According to the simulation results, we divide the spreading processes into two classes. For the first class, ρ_s is a fixed value for different simulations with the same α and β. In this case, the value of ρ_s is determined by α and β. The first class marks with red zone in figure 2. For the second class, ρ_s has different values for different simulations. In this case, the value of ρ_s is determined by the first infected node besides α and β, as figure 1 shows. The second class marks with gray zone in figure 2. The numbers on red zone represent the value of ρ_s with corresponding α and

β. e.g. When $\alpha = 3$ and $\beta = 2$, the steady infected nodes ratio is close to $\rho_s \approx 0.42037$, for different simulations.

β \ α	1	2	3	4
4	0. 32982			
3	0. 35945	0. 32928		
2	0. 38817	0. 37217	0. 42037	0. 43457
1	0. 43147	0. 41853	0. 45713	0. 4675

Figure 2. States of the spreading process with $\alpha = 1, 2, 3, 4$ and $\beta = 1, 2, 3, 4$. The numbers on red zone represent the value of ρ_s with corresponding α and β.

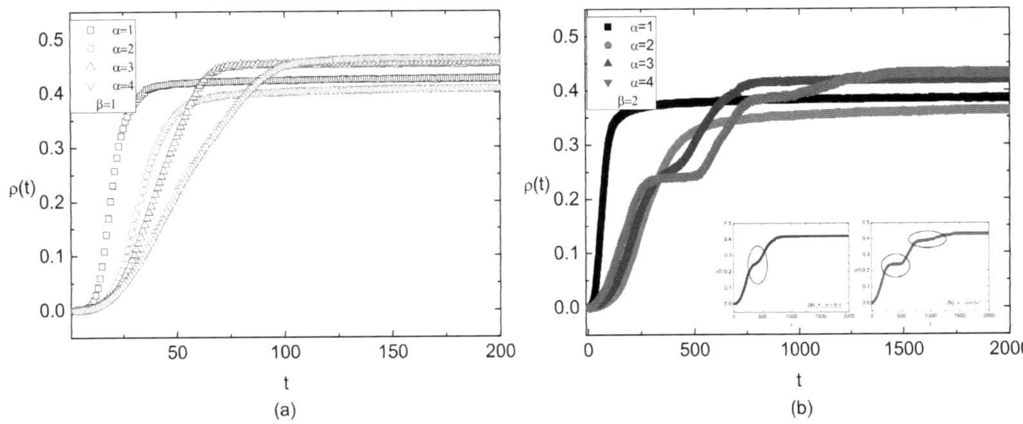

Figure 3. The infected nodes ratio $\rho(t)$ as a function of time t in two special cases, (a) $\beta = 1$ and (b) $\beta = 2$ with $\alpha = 1, 2, 3, 4$, respectively.

Figure 3 shows the evolution of infected nodes ratio $\rho(t)$ as a function of time t for different spatial driven networks and different β. According to eq. (3), spatial driven networks with different α have the same degree distribution. But in figure 3(a) and figure 3(b) the spreading behaviors are different. In S-SIS model the degree distribution is no more the main element for epidemic dynamics. We have discussed this property in Ref. [16]. Comparing figure 3(a) and figure3(b), we reveal that the curves for $\alpha = 3, \beta = 2$ and $\alpha = 4, \beta = 2$ have platform areas when $\rho(t)$ grows as a function of time t. In figure 3(b) insets, circles mark this phenomenon. $\beta = 2$ means the medium is difficult to spread disease in epidemic spreading, $\alpha = 3$ and $\alpha = 4$ means short spatial distance network. Both epidemic dynamics and network spatial structure go against disease spreading. But when disease spreads to some special nodes, disease erupts again.

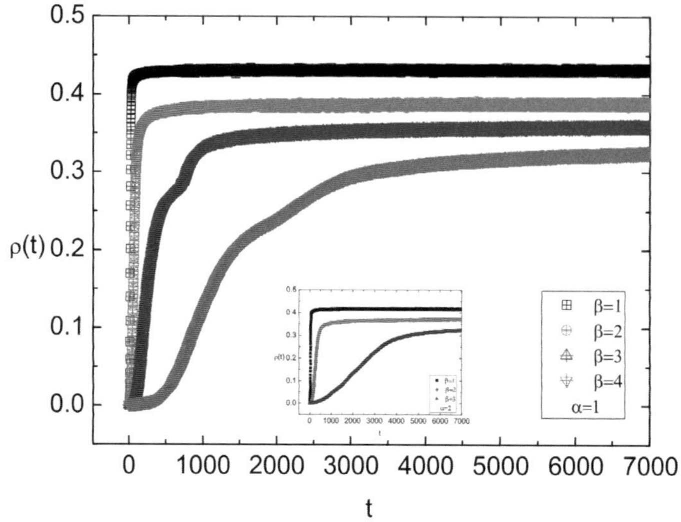

Figure 4. The infected nodes ratio $\rho(t)$ as a function of time t with $\alpha = 1$ and $\beta = 1, 2, 3, 4$. The insert is the results for $\alpha = 2$ and $\beta = 1, 2, 3$.

In figure 4, we plot $\rho(t)$ as a function of time t for different $\beta = 1, 2, 3, 4$ on spatial driven network with $\alpha = 1$. In figure 4 insert, we only plot three curves with $\beta = 1, 2, 3$ when $\alpha = 2$. Because in figure 2, we have discussed that, when $\alpha = 2$ and $\beta = 4$, the value of ρ_s is determined by the first infected node. We do not mention this case in figure 4 insert. In figure 4 and figure 4 insert, we clearly notice a great influence of infective medium on effective spreading time of disease and the steady infected nodes ratio ρ_s. The effective spreading time is the number of steps to reach to the steady state. With the increasing of β, the disease needs longer effective spreading time to reach to the steady state, ρ_s gets smaller value. According to eq. (2), for larger β, λ_{sj} gets a smaller value. Thus, these infective media have smaller probability to transmit disease, which cause smaller ρ_s and longer effective spreading time.

4. Summery

In summary, this paper presents Spatial Susceptible-Infectious-Susceptible model to discuss epidemic dynamics on spatial driven network. In this model, infective rate of a node is determined by its spatial connections and infective medium. For larger α and β, spreading steady state ρ_s is determined by the first infected node. In others, ρ_s is independent of the first infected node. With the increasing of β, epidemic behavior needs longer time to reach the steady state, but the value of ρ_s is smaller. Larger β makes infective medium go against epidemic spreading. In the case of $\beta = 2$ and $\alpha = 3, 4$, when disease spreads to some special nodes, disease erupts again. It will be another interesting work, finding properties of special nodes in S-SIS model, in the future.

Acknowledgements

This work is supported by the Research Project of Hubei Provincial Department of Education No. Q20161703.

References

[1] Bailey N T J 1975 *The Mathematical Theory of Infectious Diseases and Its Applications* New York: Hafner Press
[2] Anderson R M and May R M 1992 *Infectious Disease of Humans* Oxford: Oxford University Press
[3] Hetheote H W 2000 *SIAM Reviw* **42** 599
[4] Pastor-Satorras R and Vespignani A 2001 *Phys.Rev.Lett.* **86** 3200
[5] Pastor-Satorras R and Vespignani A 2001 *Phys.Rev.E* **63** 066117
[6] Kephart J O, White S R, and Chess D M 1993 *IEEESpectr.* **30** 20
[7] Guìmerà R, Danon L, Díaz-Guilera A, Giralt F and Arenas A 2003 *Phys.Rev.E* **68** 065103
[8] Barrat A, Barthélemy, Pastor-Satorras R and Vespignani A 2004 *Proc.Natl.Acad.Sci.* **101** 3747
[9] Masucci A P, Smith D, Crooks A and Batty 2009 *Eur.Phys.J.B* **71** 259
[10] Solé, Rosas-Casals, Corominas-Murtra B and Valverde S 2008 *Phys.Rev.E* **77** 026102
[11] Liben-Nowell D, Novak J, Kumar R, Raghavan P and Tomkins A 2005 *Proc.Natl.Acad.Sci.* **102** 11623
[12] Kosmidis K, Havlin S and Bunde A 2008 *Europhys.Lett.* **82** 48005
[13] Hui Z, Li W, Cai X, Greneche J M and Wang Q A 2013 *PhysicaA* **392** 1909
[14] Ozik J, Hunt B R and Ott E 2004 *Phys.Rev.E* **69** 026108
[15] Barthélemy M 2011 *Phys.Rep.* **499** 1
[16] Hui Z, Cai X, Greneche J M and Wang Q A 2012 *Int.J.Mod.Phys.C* **23** 1250082
[17] Guo W P, Li X and Wang X F 2007 *PhysicaA* **380** 684
[18] Sun P, Cao X B, Du W B and Chen C L 2010 *Phys.Pro.* **3** 1811
[19] Sims D W, Metcalfe J D *et.al.* 2008 *Nature* **451** 1098
[20] Brockmann D, Hufnagel L and Geisel T 2006 *Nature* **439** 462
[21] Gonzalez M C, Hidalgo A and Barabasi A L 2008 *Nature* **453** 779

IOP Conf. Series: Journal of Physics: Conf. Series **1113** (2018) 012007 doi:10.1088/1742-6596/1113/1/012007

The principle of least effort and Zipf distribution

Yueying Zhu[1,2], Benwei Zhang[2], Qiuping A. Wang[3,4], Wei Li[2,5], Xu Cai[2]

[1] Research Centre of nonlinear Science, Wuhan Textile University, Wuhan, Hubei, 430073, China
[2] Key Laboratory of Quark & Lepton Physics (MOE) and Institute of Particle Physics, Central China Normal University, Wuhan, 430079, China
[3] IMMM, UMR CNRS 6283, Le Mans Université, 72085 Le Mans, France
[4] School of Hautes Etudes d'Ingenieur, Yncrea, 59014 Lille, France
[5] Max-Planck Institute for Mathematics in the Sciences, Inselst. 22, 04103 Leipzig, Germany.

E-mail: zhuyy@mails.ccnu.edu.cn

Abstract. "Each individual will adopt a course of action that will involve the expenditure of the probably least average of his work." This statement was named "the principle of least effort". The principle of least effort is often known as a "deterministic description of human behavior". In this paper, we present a brief introduction of this principle. Applications of the principle in different fields are also summarized. As the principle of least effort is proposed by Zipf, it is also called Zipf's law. We then discuss the correlation between three widely considered distributions: Zipf distribution, Pareto distribution and probability distribution. With empirical investigations, it is often stated that, most social behaviors are controlled by the pure Zipf's law that corresponds to the Zipf distribution of exponent -1. We summarily present the discovery of Zipf's law in different social behaviors. Some empirical studies are also given as examples, verifying that, in most countries, the distribution of city size by population follows Zipf's law, and the exponent of Zipf distribution of individual income is about -0.5, the same as Zipf predicted in theory.

1. Introduction

The principle of least effort was first discovered in 1894 by a French philosopher: Guillaume Ferrero. He discussed this principle in his article entitled "L'inertie mentale et la loi du moindre effort" [1]. However, until 1949, the principle was proposed by George Kingsley Zipf, an American professor of philology at Harvard University, in his book *"Human Behavior and the Principle of Least Effort"* [2]. Zipf theorised that the distribution of word use was due to tendency to communicate efficiently with least effort. Hence, the principle of least effort is also known as Zipf's Law. Based on the principle of least effort, it is human nature to want the greatest outcome at the least amount of work. And Zipf showed that useful behaviors were performed frequently. Frequent behaviors became quicker and easier to perform over time. With this phenomenon, it is presented that people often chose their entire behavior along the direction of minimizing the effort. Basically, Zipf's law describes people's social behavior in space.

Zipf studied the least effort of individual behavior and of collective behavior, separately. Regarding individual behavior, he statistically analyzed, for example, the speech, words and their meaning, the verbalizations of children. For the collective behavior, he mainly focused on the studies of economy of human social behavior, such as the economy of geography, the distribution of economic power and social status, the distribution of prestige symbols and vogues. In both of these two parts, Zipf began with the empiric aspect, presenting a large number of observations from a truly wide range of living phenomena. And then, he gave the underlying theoretic analysis, attempting to rationalize different kinds of empiric laws in terms of a single uniform principle. Therefore, the principle of least effort is a theory which could be well understood from both empirical and theoretical views.

Actually, the topic of effort first aroused the interest of some experimentalists. In 1930, J. A. Gengerelli published some results of a series of experiments performed with blinded and normal white rats. These rats were used to help determine the nature of the path that animals would eventually select from an indefinite number of possible paths leading to food [3]. Experimental results suggested that in practically all cases that the path finally chosen by animals (both normal and blinded ones) was the path of "least effort," namely, the path of minimal distance. Two years later, L. S. Tsai stated that "Among several alternatives of behavior leading to equivalent satisfaction of some potent organic need, the animal, within the limits of its discriminative ability, tends finally to select that which involves the least expenditure of energy" [4]. And also, in 1937, R. H. Waters declared that "Thus, Theseus, after slaying the minotaur, found his way out of the labyrinth and to his loved one by following the string which he had carried with him into the labyrinth. Perhaps, this was not the most direct route in terms of distance, time, or effort, but it was the only sure way he had of escaping. Likewise our rats found that by sticking to the outside pathways they more readily achieved the goal" [5].

Since the principle of least effort was proposed by Zipf, many scholars are attracted to study it and to apply it to different fields. In the next section, we will discuss applications of the principle of least effort in different fields.

2. The principle of least effort in different fields

2.1. The field of information retrieval

The principle of least effort is exceptionally important in designing libraries and researching in the context of modern library. User's desire to find information quickly and easily is often the primary consideration in a library design. In library literature, "least effort" is restated as Mooers' law, an empirical observation of behavior provided by an American computer scientist Calvin Mooers in 1959 [6]. More commonly, Mooers's law is considered to be a derivation of the principle of least effort. With Mooers' law, it is stated that an information retrieval system will tend not to be used whenever it is more painful and troublesome for a user to have information than for him not to have it.

In 1987, T. Mann classified the principle of least effort as one of several principles controlling information seeking behavior [7]. He emphasized seven different research methods helping individuals get further into a subject more quickly, and with less wasted effort. Furthermore, E. G. Bierbaum also declared that "least effort" was one scholar's suggestion for the uniform principle needed in research and practice, library and information science [8]. She stated "No other principle underlies as much of library and information science. Least effort explains the one-look-up reader, staff resistance to automation, the reliance of the scientist in colleagues rather than collections, and the rapid acceptance of CD-ROM compared to microfilm".

T. E. Chrzastowski, a professor of library administration, discussed in 1995 whether the workstations, aided by the principle of least effort, had changed the nature of how research was performed in academic libraries [9]. This discussion was proposed based on an investigation about library workstation popularity and the principle of least effort. With empirical data, she first analyzed the impacts of IBIS databases on the UIUC Chemistry library and then the journal use in the UIUC Chemistry library. Empirical results supported the principle of least effort and also showed that "least effort" method was often considered by many patrons as a suitable model for academic library research.

In 2004, the principle of least effort was further explored by Z. Liu and Z. Yang [10]. They provided a self-administered questionnaire to study the principal individual and environmental factors influencing a student's decision process of selecting and using their information sources. Survey results say that reasons given by respondents for selecting and using primary information sources show their strong preference for fast and easy information retrieval. This phenomenon suggests that the principle of least effort also controls the respondent's selection and use of information sources.

2.2. The field of human behavior

Generally speaking, interpretation of palaeoenvironmental changes is a key mission of researchers from archaeology, botany, geography and many other relevant disciplines. The principle of least effort also owns its specified scope of application in this field. Employing the principle, scholars A. Scholtz and M. L. Tusenius respectively provided a functional interpretation of charcoal data sets in 1986 [11, 12].

In 1992, C. M. Shackleton and F. Prins further discussed the applicability of the principle of least effort in explaining palaeoclimatic data [13]. They proposed a conceptual model to determine whether the principle of least effort was applicable to a given situation. In the conceptual model, four generalized areas used for inhabitant's collection of fuelwood were described, occurring in chronological order. Area one regards the situation of plenty of fuelwood, both dead and live, with a considerable diversity of species available. Area two considers the situation of a medium abundance of dry wood, declining selectivity and maximum effort. Areas three and four take into account the conditions of low wood abundance, little selectivity, minimal effort and demand for wood far exceeding supply. Experimental results display that the principle of least effort is only applicable to the situation of area three. This conceptual model can not only help identify when the principle of least effort is appropriate for interpreting some data sets, but also be a valuable aid for understanding the past human behavior.

In 2002, R. F. Cancho and R. V. Sole illustrated a hypothesis of Zipf on the principle of least effort [14]. They aimed to provide new theoretical insights into the absence of intermediate stages between animal communication and language. Beginning with this idea and purpose, Cancho and Sole established a simple form of language game employing a mathematical model that involves a set of signals and objects. With the game, they studied the problem of compromise between speaker and hearer needs. Results strongly display that Zipf's law is a hallmark of symbolic reference but not a meaningless feature.

2.3. The field of animal behavior

In animal ethology, an animal often wants to achieve the most energy with the lowest cost while foraging, so as to maximize the fitness. The optimal foraging theory is a model that helps predict the best strategy with which an animal can achieve the goal stated above. This theory is well known as the most essential theory helping predict how an animal behaves while collecting food. So, the optimal foraging theory can be considered as a derivative of the principle of least effort. The optimal foraging theory assumes that the most economically advantageous foraging pattern will be selected by a species through natural selection that has achieved optimal allocation of time and energy expenditures[15]. In 1984, A. Kacelnik found that starlings could maximize net energy gain per unit time [16]. J. R. Krebs and N. B. Davies also stated in 1989 that, with maximal energy efficiency, the bees were able to avoid expending too much energy per trip and to live long enough to maximize their lifetime productivity for their hive [17].

However, a long term assumption regarding livestock trails in grazing procedure said that livestock often established their pathways of least resistance between frequented portions of their pastures [18, 19]. This assumption was proposed by considering only a component of MacArthur and Pianka's optimum foraging theory, stating that animals were expected to minimize energy expenditures while maximizing energy gains in their daily endeavors. It had never been verified because of the difficulty in quantification of foraging activities and of optimal energy in practice. Until 2000, with the help of Geographic Information Systems (GIS) and Global Positioning Systems (GPS), Ganskopp *et al.* built a method to quantify the extent and characteristics of cattle trails in three large pastures, and indirectly verified three hypotheses [20]. They mapped cattle trails in three 800+ ha pastures containing global positioning units. Then they used GIS to quantify characteristics of both trails and the landscape, and also to plot least-effort pathways connecting water sources and distant points on selected trails in each pasture. Furthermore, the assumption that "cattle develop lest-effort routes of travel in rugged terrain by comparing the characteristics of cattle trails and least-effort pathways" was also testified.

3. Zipf distribution

3.1. The relationship between three kinds of distribution

In the field of empirical investigation, three widely considered distributions include the probability distribution, Pareto distribution, and Zipf distribution. Many man made and naturally occurring phenomena, such as city sizes, incomes, word frequencies, and earthquake magnitudes, are distributed according to a power-law distribution[21]. This power-law behavior is often presented by the probability distribution that describes the probability of occurrence of an event during the whole system. Zipf distribution focuses on the correlation between the frequency of occurrence of an event and the rank of the underlying event. Pareto distribution is a cumulative distribution, illustrating the probability that a person has an income no less than a given number. Though these three distributions are established in different aspects, they refer to the same thing. In this section, we will discuss the correlation between them. Specifically, we mainly focus on the derivation of both Pareto and probability distributions from a Zipf one.

We take the distribution of individual income as example. Zipf's law states that the size of income for individual people is inversely proportional to the rank of this income in decreasing order. Mathematically, Zipf's law is formulated as

$$I_r \propto C_1 * r^{-1/2}, \tag{1}$$

where I_r is the income of rank r in decreasing order, and C_1 the size of income with rank 1. For convenience, we label the exponent in Zipf distribution by a general variable α, that is

$$I_r = C_1 * r^{-\alpha}. \tag{2}$$

The rth income owns size $C_1 * r^{-\alpha}$, meaning that there are r kinds of income with size no less than $C_1 * r^{-\alpha}$. With this description, the probability that an income is no less than $C_1 * r^{-\alpha}$ is proportional to r, expressed as

$$P(X \geq C_1 * r^{-\alpha}) = C_2 * r. \tag{3}$$

Setting $x = C_1 * r^{-\alpha}$, we can get $r = [\frac{x}{C_1}]^{-\frac{1}{\alpha}}$. Pareto distribution regarding the same event is then represented as

$$P(X \geq x) = C_2 * [\frac{x}{C_1}]^{-\frac{1}{\alpha}}. \tag{4}$$

With Pareto distribution, we can also directly get the cumulative probability distribution, that is

$$\begin{aligned} P_r(X < x) &= 1 - P(X \geq x) & (5)\\ &= 1 - C_2 * [\frac{x}{C_1}]^{-\frac{1}{\alpha}}. & (6) \end{aligned}$$

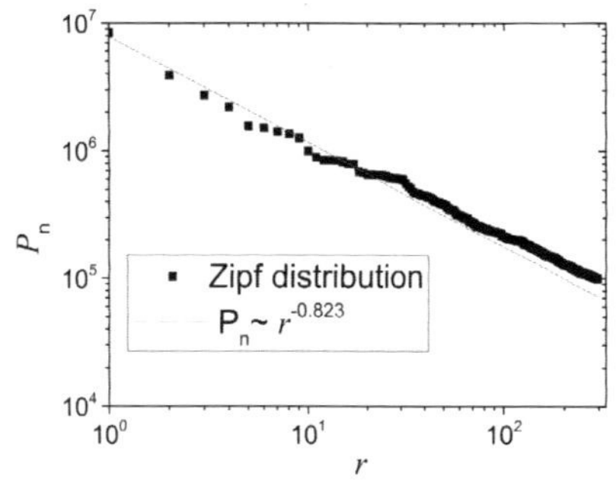

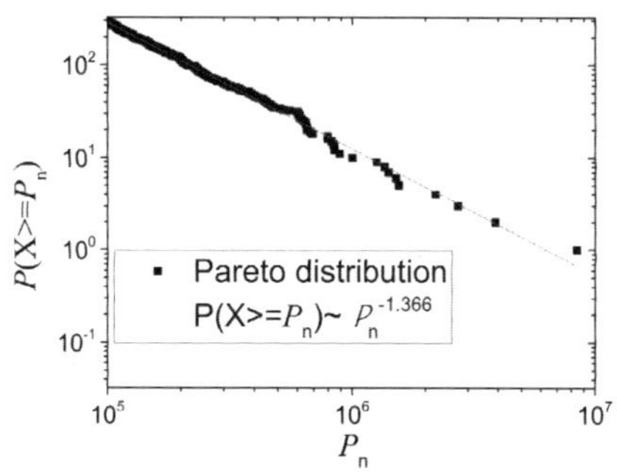

Figure 1. Zipf and Pareto distributions of city size by population in United States for 2013.

Taking the first derivative of the above equation with respect to x, we then have the probability distribution:

$$P_r(X = x) = \frac{dP_r(X < x)}{dx} \tag{7}$$

$$= \frac{C_2}{\alpha} * C_1^{\frac{1}{\alpha}} * x^{-(\frac{1}{\alpha}+1)}. \tag{8}$$

According to the above discussion, it is noticed that a Zipf frequency-rank distribution of exponent α corresponds to a Pareto cumulative distribution of exponent $\frac{1}{\alpha}$ and a power-law probability distribution of exponent $(\frac{1}{\alpha}+1)$. X. Gabaix and Y. M. Ioannides illustrated in 2004, by Monte Carlo simulations, that it was considered as a successful application of Zipf's law when the exponent of Pareto distribution was between 0.8 and 1.2 [22].

Figure 1 presents a simple empirical investigation regarding both Zipf and Pareto distributions of American city size by population in 2013. It is illustrated that the exponent of Pareto distribution is about -1.366, quite close to the reciprocal of the exponent of Zipf distribution -0.823. This statistical phenomenon directly provides the empirical evidence for the correlation between exponents of Zipf and Pareto distributions.

3.2. Alternative expressions of Zipf's law

Zipf's law can be generalized by an approximation on the relationship between rank and frequency

$$r^{\alpha} * f = C, \tag{9}$$

where r is the rank of a word-type in decreasing order of frequency, f the frequency of occurrence of the corresponding word in a given text, and C a constant. C, depending upon the underlying text, is often about one-tenth of the size of the text (the total number of running words). When $\alpha = 1$, Eq. (9) is well known as the pure form of Zipf's law:

$$f = C * r^{-1}. \tag{10}$$

Pure Zipf's law states that the size of the rth largest occurrence of an event is inversely proportional to its rank.

In 1954, B. Mandelbrot proposed a further refinement of Zipf's law:

$$f = C * (r + \rho)^{-B}, \tag{11}$$

in which r is the rank of a word, f the frequency, C, ρ and B the constants dependent upon the underlying text[23]. This expression forms the basis of statistical LNRE (large number of rare events) models and provides a better fit to low rank but high frequency words. H. P. Edmundson developed a new 3-parameter expression in 1972 considering the relationship between frequency and rank:

$$f(r; c, b, a) = c * (r + a)^{-b}, c > 0, b > 0, a \geq 0, \tag{12}$$

where a, b and c are constants[24].

Except for the distribution of words by frequency in a given text, the behavior of Zipf's law is also exhibited in many other aspects: the distribution of city size by population (as presented above), the distribution of individual income, the distribution of scientist by the number of published papers, etc.

3.3. Zipf's law in different aspects

3.3.1. The distribution of city size The distribution of city size by population has been a long-standing topic of interest since the last century. In 1913, F. Auerbach, a German physicist, launched an initial interest in the distribution of city size [25]. With empirical data, he found there was a universal relationship connecting the city's population and city's rank in United States and five European countries. This relationship can be denoted by,

$$P_i * r_i = A, \tag{13}$$

where A is a constant, P_i the average population of cities in size-class i, and r_i the rank of class i in the order of size decreasing.

A. J. Lotka, a US scholar, found empirically in 1925 a better fit function for the first 100 largest cities in United States in 1920 [26]. The function is expressed as

$$P_i * r_i^{0.93} = 5,000,000. \tag{14}$$

In 1940, Zipf analyzed the rank-frequency distribution of the first 100 largest Metropolitan Districts in United States in 1940, according to the Sixteenth Census. He discovered that the slope of the underlying distribution was -0.98 (see Fig. 9-2 in Ref. [2]). Zipf's law in the distribution of city size by population suggests that the city with the largest population in any country is generally twice as large as the next-biggest, and three times as large as the third biggest, and so on. Incredibly, Zipf's law in city size distribution has always held true for most countries in the world.

Furthermore, K. T. Rosen and M. Resnick classically studied in 1980 the rank-size distribution of cities of 44 countries encompassing both developing and developed nations, using 1970 census data [27]. With their results, the first 50 largest cities in most countries can be predicted by a Pareto distribution of the form:

$$R = AS^{-a}, \tag{15}$$

where R is the number of cities with population S or more, A a constant, S the population of city, and a the Pareto exponent. For 44 countries under their investigation, exponent a ranges from 0.809 to 1.963 and has mean value 1.136. And for 32 of them, the Pareto exponent is greater than one. This discovery suggests that the population in most countries is more evenly distributed than that could be predicted by the pure Zipf's law which corresponds to a Pareto distribution of exponent 1. Similar results were also provided by K. T. Soo in 2005 who assessed the empirical validity of Zipf's law in the distribution of city size using new data of 73 countries and two estimation methods [28]. In 2007, he also performed a test of Zipf's law in the rank-size distribution of cities, based on five years population censuses from Malaysian (1957, 1970, 1980, 1991 and 2000) [29]. Meanwhile, the factors that possibly influenced the growth of a city in

Malaysia were also explored. Soo found that Zipf's law held for the size-rank distribution of cities in Malaysia in 1957. But since then, cities were more unequal in size than that would be predicted by Zipf's law. Furthermore, he also stated that a city growth was negatively related to the city size. This statement is against Gibrat's law, a rule defined by R. Gibrat, stating that the growth rate of a city is independent of the city size [30, 31].

In 2012, M. Cristelli *et al.* also found that Zipf's law held approximately for the distribution of city size in each European country (France, Italy, Germany, etc). While for the aggregated data in European union, Zipf's law completely failed. They declared that "In fact, historically, the geographic level for Europe, at which an integrated evolution is observed, is the national state, while in the US, the whole confederation, not each independent state, has collectively and organically evolved towards a distribution of cities that follows Zipf's law. From this perspective, the US is an organic, integrated economic federation, while the EU has not yet become so, and shows little convergence to such an economic unit" [32]. This would seem to support the idea that Zipf's law is a response to economic conditions, since it works only if you compare cities that are connected economically in a country.

With theoretical analysis and empirical studies, Y. Chen found that Zipf's law was closely related to the hierarchical scaling law [33]. Beginning with a general form of Zipf's law of exponent q, the author firstly defined a self-similar hierarchy of cities based on q-sequence. And then he deduced theoretically the hierarchical scaling law of cities, and observed that the exponent of rank-size scaling law was the reciprocal of that of size-number scaling law. Empirical data from both America and China showed the existence of Zipf's law in the rank-size distribution of cities with exponent -0.738 and -0.889, respectively. The exponent of the number-size distribution was -1.364 for America and -1.193 for China.

In 2013, S. Li and D. Sui studied the rank-size distribution of China's urban system based on empirical data in the period of 24 years: from 1984 to 2008 [34]. They displayed that the upper tails of rank-size distributions of Chinese cities followed a power-law function of slope slightly less than -1. This characteristic suggested that Chinese cities were more evenly distributed than that predicted by Zipf's law. A similar trend was also found in many other countries [35]

J. Luckstead and S. Devadoss provided in 2014 a comparison and examination of size distributions of Chinese and India cities from 1950 to 2010, using log-normal, Pareto, and general Pareto distributions. Results provide that large number of cities from China and India have similar trends: the rank-size distribution of cities is log-normal in the early periods but power-law in 2010. Furthermore, in both 2000 and 2010, the distribution of city size in India is controlled by pure Zipf's law, different from the situation in China [36].

An empirical investigation regarding the distribution of city size is also presented here, see Fig. 2$(a-e)$. We mainly focus on the first 295 largest cities in United States in both 2010 and 2013, major cities in Germany of more than 100,000 inhabitants in 2010, major cities in France of more than 75,000 inhabitants in 1999, major cities in China of more than 155,540 inhabitants in 2010, and major cities in Japan of more than 202,283 inhabitants in 2010. Results reveal that Zipf distributions of the size (denoted by population) of cities under investigation all behave obviously power-law distribution:

$$P \sim r^{-\alpha}, \tag{16}$$

P labels the population of a city and r the underlying city's rank. For United States, Germany, France and Japan, the exponent α of the distribution of city size by population values in sequence -0.823/-0.827, -0.809, -0.996 and -0.846. These values are quite close to -1, suggesting the distribution of city size by population in countries mentioned above is approximately controlled by the pure Zipf's law. In China, however, the exponent α is -0.590, diverging from the value -1 indicating the pure Zipf's law. This phenomenon sates again that the distribution of population in China is more homogeneous than that predicted by pure Zipf law. This might be because China owns the largest population, compared to other countries. More similar distributions from other countries could be found in Ref. [2].

3.3.2. The distribution of firm size As early as 1949, Zipf illustrated the rank-frequency (the number of wage earners) distribution of manufactures in United States. With data collected in 1939 and manufactures ranked in the order of their decreasing number of wage earners, he showed the slope of the underlying distribution was -2/3 (Fig. 9-8 in Ref. [2]). Zipf also studied the rank-frequency distribution of corporation assets in United States, based on data dating from 1931 to 1936. Results exhibit that corporation assets act as a power-law function of corporation's rank with exponent close to -1 [2].

It has often been observed that the upper tail of the size distribution of firms resembles a Pareto distribution. However, people does not know how to explain this kind of distribution by economic theory. In 1958, H. A. Simon and C. P. Bonini developed a stochastic model of the growth process of firms on the basis of two assumptions [37]. Simulation results provided by the model were quite consistent with empirical data. With their assumptions, Yule distribution would be the steady-state distribution of the same process. Regarding the distribution of firm size, Yule distribution in the upper tail can be approximated by the Pareto distribution: $f(s) \propto s^{-(\rho+1)}$ as $s \to \infty$. Simon and Bonini showed that parameter ρ here was represented by

$$\rho = \frac{1}{1 - g/G}, \tag{17}$$

in which, G is the net growth during some specified periods of assets of all firms in an industry, and g the part of net growth attributable to new firms (firms that have reached the minimum size during the period).

In 1995, with data of American firms from Compustat, M. H. R. Stanely *et al* demonstrated using Zipf plot that the upper tail of the distribution of firm size was too thin, relative to the log normal, rather than too fat [38]. They reported that the distribution of American firm size could be approximated as a log-normal distribution. This discovery, however, could not explain the general property of firms as the empirical data they used were unrepresentative of overall population of American firms. Regarding this limitation, R. L. Axtell studied in 2001 the distribution of American firm size,with the entire population considered, based on the data combined from Census/SBA and Compustat [39]. With "non-employee" firms (of size 0) neglected, he found the distribution of firm size by the number of employees followed Zipf's law of slope -1.008. And the slope would be -0.963 if those firms of size 0 were included. Based on a large number of empirical results, researchers also stated that "the Zipf distribution of firm size is robust across varying definition of 'size', and so too it is quantitatively invariant over time" [40].

In view of the fact that most of researches only focus on the economy in developed countries, J. Zhang *et al* analyzed in 2009 the data of top 500 Chinese firms from 2002 to 2007 [41]. China is commonly regarded as the biggest developing country in the world. Her rapid growth in economy always attracts the interest of many researchers from different fields. With empirical investigation, Zhang *et al* revealed that the revenues and ranks of Chinese firms exactly obeyed the pure Zipf's law, with exponent being -1 for each year under consideration. Furthermore, they offered an explanation of this characteristic using a simple economic model, namely AK model. The model is usually applied to describe how the revenue of a firm depends upon its capital and technology level by neglecting the impact of human resource [42]. AK model, however, can satisfactorily represent the formation process of firm size only in China. This is because human resource is more abundant and much cheaper for Chinese firms, compared to other aspects of consumption [43].

3.4. The distribution of individual income

The distribution of individual income has always been a central concern in economic theory and policy. We have discussed above the rank-frequency distribution of firm size which is analogous to the distribution of group income. Zipf indicated theoretically that the slope of the rank-frequency distribution of firm size was -1, in good agreement with empirical results.

However, for the rank-frequency distribution of individual income, the slope should equal -0.5 except the case of the settlement of liability claims against automobile insurance companies whose slope would often rise to 1. This value was predicted theoretically by Zipf in 1949 [44]. Based on individual income data samples from seven different countries including the United Kingdom, France, Finland, etc, he also illustrated some empirical results of the rank-frequency distribution of individual income. All these results exhibit a very good verification of the theoretical prediction of slope -0.5, independent of the country and the year that data samples come from.

But differently, H. T. Davis claimed as early as 1941 that the slope of Pareto distribution of individual income was -1.5, according to a very large amount of empirical data [45]. We have discussed before the reciprocal relationship between the exponent of Pareto cumulative distribution and that of Zipf rank-frequency distribution. Based on Davis's results, the slope of rank-frequency distribution of individual income in double logarithmic coordinates was -2/3. Davis also argued that the slope would tend to fall in good times and to increase in bad times.

A simple empirical investigation is also given as an example here, regarding the Pareto distribution of individual income in United States for 2010. The underlying result is presented in Fig. 2(f), indicating the exponent of Pareto distribution is close to -2. This suggests the slope of Zipf rank-frequency distribution in double logarithmic coordinates is about -0.5, the same as Zipf predicted in theory.

4. Conclusions

We first present a brief introduction of the principle of least effort. Applications of this principle in the field of information retrieval, human behavior and animal behavior are also summarized. Furthermore, we also discuss the correlation between three widely used distributions: the Zipf distribution, Pareto distribution and probability distribution. Results suggest that both Pareto and probability distributions can be derived from Zipf distribution. A Zipf distribution of exponent α is analogous to the Pareto distribution of exponent $\frac{1}{\alpha}$, and the probability distribution of exponent $(1+\frac{1}{\alpha})$. Zipf distribution considers the correlation between the frequency of occurrence of an event and the rank of the underlying event. With empirical investigations, it is widely stated that Zipf distribution acts as power-law decay function in most social behaviors. And the corresponding exponent is always close to -1, namely the pure Zipf's law. Regarding individual income, however, the exponent of Zipf distribution is close to -0.5, predicted in theory and also verified by empirical investigation by Zipf. We summary discuss the discovery of Zipf's law in different aspects. Some empirical investigations are also provided as examples in this paper. Results verify again that, in different countries, the distribution of city size by population is controlled by Zipf's law of slope close to -1. The slope of Zipf distribution of individual income is about -0.5, the same as Zipf predicted in theory.

Acknowledgments

This work was partly supported by the National Natural Science Foundation of China under Grant Nos. 11435004 and 11521064, and by the Programme of Introducing Talents of Discipline to Universities under Grant No. B08033.

References

[1] Ferrero G 1894 *Revue Philosophique de la France et de l'étranger* **37** 169C182

IOP Conf. Series: Journal of Physics: Conf. Series **1113** (2018) 012007 doi:10.1088/1742-6596/1113/1/012007

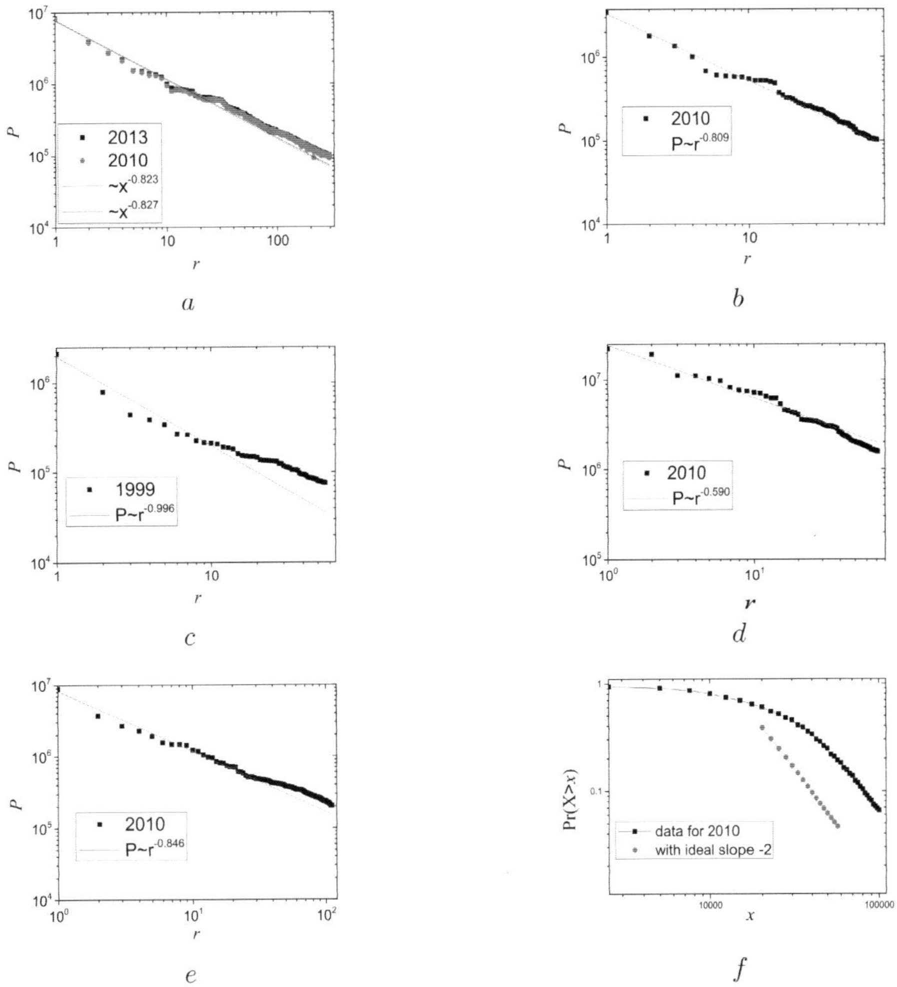

Figure 2. The Zipf distribution of city size by population ($a - e$) and Pareto distribution of individual income (f). a) Results for American cities. Black and red symbols indicate the data of April 1, 2010 enumerated by the 2010 United States Census and of July 1, 2013 estimated by the United States Census Bureau, respectively. b) Results for German cities of more than 100,000 inhabitants in 2010. c) Results for major cities in France of more than 75,000 inhabitants in 1999. d) Results for major cities in China of more than 155,540 inhabitants in 2010. e) Results for major cities in Japan of more than 202,283 inhabitants in 2010. f) The Pareto distribution of individual income in United States in 2010 (line+black squares). Line+red circles has an ideal slope of -2.

[2] Zipf G K 1949 *Human behavior and the principle of least effort: An introduction to human ecology* (Cambridge: Addison-Wes-ley)
[3] Gengerelli J A 1930 *J. Comp. Psychol.* **11** 193-236
[4] Tsai T S 1932 *The laws of Minimum Effort and Maximum satisfaction in Animal Behavior* Monograph of the National Institute of Psychology
[5] Waters R H 1937 *J. Gen. Psychol.* **16** 3-20
[6] Mooers C N 1959 *Bull. Am. Soci. Inf. Sci. & Tech.* **23** 22-23
[7] Mann T 1987 *A guide to library research methods* (Oxford: Oxford University Press)
[8] Bierbaum E G 1990 *American Libraries* **21** 18-19
[9] Chrzastowski T E 1995 *J. Am. Soc. Inf. Sci. 46* 638-641
[10] Liu Z, and Yang Z 2004 *The Journal of Academic Librarianship* **30** 24-35
[11] Scholtz A 1986 *Palynological and palaeobotanical studies in the southern Cape* M. A. thesis (Stellenbosch: University of Stellenbosch)
[12] Tusenius M L 1986 *The study of charcoal from some southern African archaeological contexts* M. A. thesis (Stellenbosch: University of Stellenbosch)
[13] Shackleton C M and Prinsh F 1992 *J. Archaeolog. Sci.* **19** 631-637
[14] Cancho R F and Sole R V 2002 *PNAS* **100** 788-791
[15] Macarthur R H and Pianka E R 1966 *Am. Nat.* **100** 603-609
[16] Kacelnik A 1984 *J. Anim. Ecol.* **53** 283-299
[17] Krebs J R and Davies N B 1989 *An Introduction to Behavioral Ecology* 4th ed. (Oxford: Blackwell Scientific Publications)
[18] Weaver J E and Tomanek G W 1951 *Nebr. Conserv. Bull.* **31** 1-82
[19] Arnold G W and Dudzinski M L 1978 *Developments in Animal and Veterinary Science: Ethology of Free-ranging Domestic Animals* (New York: Elsevier)
[20] Ganskopp D *et al* 2000 *Appl. Anim. Behav. Sci.* **68** 179-190
[21] Adamic L A 2000 *Zipf, Power-laws, and Pareto-A ranking tutorial* http://ginger. hpl. hp. com/shl/papers/rankin-g/ranking.html.
[22] Gabaix X and Ioannides Y M 2003 *Discussion Papers* **4** 2341-78
[23] Mandelbrot B 1954 *Word* **10** 1-27, 424-25
[24] Edmundson H P 1972 *The Rank Hypothesis: A Statistical Relation between Rank and Frequency* Technical report TR-186. (College Park: Computer Science Center, University of Maryland)
[25] Auerbach F 1913 *Petermanns Geographische Mitteilungen* **59** 74-76
[26] Lotka A J 1925 *Elements of physical biology* (Baltimore: Williams and Wilkins)
[27] Rossen K T and Resnick M 1980 *J. Urban Econ.* **8** 165-186
[28] Soo K T 2005 *Reg. Sci. Urban Econ.* **35** 239-263
[29] Soo K T 2007 *Urban Studies* **44** 1-14
[30] Gibrat R 1931 *Les inégalités économiques* (Paris: Recueil Sirey)
[31] Sutton J 1997 *J. Econ. Literature* **XXXV** 40-59
[32] Cristelli M, Batty M and Pietronero L 2012 *Scientific Reports* **2** 812
[33] Chen Y 2012 *Physica A* **391** 3285-99
[34] Li S and Sui D 2013 *GeoJournal* **78** 615-25
[35] Carroll G R 1982 *Progress in Human geography* **6** 1-43
[36] Luckstead J and Devadoss S 2014 *Econ. Lett.* **124** 290-95
[37] Simon H A and Bonini C P 1958 *Am. Econ. Rev.* **48** 607-17
[38] Stanely M H R *et al.* 1995 *Econ. Lett.* **49** 453-57
[39] Axtell R L 2001 *Science* **293** 1818-20
[40] Ijiri Y and Simon H A 1935 *J. Roy. Stat. Soc.* **140** 219-23
[41] Zhang J, Chen Q and Wang Y 2009 *Physica A* **388** 2020-24
[42] Romer P 1986 *J. Polit. Econ.* **94** 1002-37
[43] Su Y and Xu X 2002 *Econ. Res. J.* **11** 3-11
[44] Zipf G K 1949 *Human behavior and the principle of least effort* (Addison-Wesley Press) 496-500
[45] Davis H T 1941 *The Theory of Econometrics* (Bloomington: The Principia press)

IOP Conf. Series: Journal of Physics: Conf. Series **1113** (2018) 012008 doi:10.1088/1742-6596/1113/1/012008

Mean-field theory for the Nagel-Schreckenberg model with overtaking strategy

Zhu Su[1*], Weibing Deng[2], Jihui Han[3], Wei Li [2], Xu Cai[2]

[1]National Engineering Laboratory for Technology of Big Data Applications in Education, Central China Normal University, Wuhan, P. R. China

[2]Complexity Science Center, Institute of Particle Physics, Central China Normal University, Wuhan, P. R. China

[3]School of Computer and Communication Engineering, Zhengzhou University of Light Industry, Zhengzhou, P.R. China

E-mail: suz@mail.ccnu.edu.cn

Abstract. Based on the Nagel-Schreckenberg (NS) model with periodic boundary conditions, a modified model considered overtaking strategy (NSOS) has been proposed [1, 2]. In this paper, we focus on the theoretical analysis of traffic flow for NSOS model by using mean-field method. In the special case of $v_{max} = 1$ where vehicles can not overtake preceding ones, the features of stationary state can be obtained exactly. However, in the case of $v_{max} > 1$ where overtaking happens, some approximative methods have to be took into account. The main results are that we find the reason why traffic flow is increased in the regime where densities exceed the maximum flow density, and the influence of traffic flow on the transition density is dominated by the braking probability p.

1. Introduction

Various dynamical models [3, 4] have been proposed to explain the complexity phenomena generated by traffic flow. From the microscopic point of view, the vehicular traffic system can be regarded as being composed of interacting particles driven far from equilibrium, each individual vehicle is represented by a particle, and the way that they influence others' movement is treated as the interactions among particles. Therefore, vehicular traffic offers the possibility to study various fundamental aspects of the dynamics of non-equilibrium systems which are of interest in statistical physics.

During the last two decades, cellular automata (CA) [5] have obtained popularity due to their simplicity and their ability to simulate large networks. One of the early CA based on traffic models is the NS model [6] developed by Nagel and Schreckenberg, and then a large amount of improved versions have been proposed by imposing some conditions on NS model to make it more realistic. In the NS model, the road is divided into sites, each site can be either empty or occupied, and all the space, time and velocities are discrete. Under periodic boundary conditions, the number of vehicles on the road remains unchanged. The state of vehicle is characterized by an internal parameter v ($v = 0, \ldots, v_{max}$), where v_{max} is the maximum velocity. To obtain the system's state in the next time, one could adopt the following rules to all vehicles at the same time (parallel dynamics): (1) The first step is an acceleration process, if a vehicle's velocity (v)

is lower than the speed limit (v_{max}), its velocity is advanced by one. (2) The second step is designed to avoid accidents, if two adjacent vehicles have h empty sites between each other, and the following vehicle has a speed larger than h, then its velocity is reduced to h. (3) The third step is considered random braking, a noise with probability p to reduce the velocity of a moving vehicle ($v > 0$) to $v - 1$. (4) The last step deals with the vehicle' movement, which enables the position of a vehicle to be advanced by its speed v.

Considered the overtaking case, the NSOS model has been proposed [1, 2], where every vehicle could be an overtaking one with probability q at each time step. The model exists due to the following facts: (1) Overtaking obviously happens a lot, especially when the preceding vehicles move quite slowly. (2) The overtaking vehicles are used to go back to the original lane once overtake successfully. (3) When the preceding vehicles are overtaken, they would slow down rather than accelerate.

Several approaches for analytical descriptions of the NS model have been studied [7, 8, 9, 10, 11, 12, 13], and they yield the exact solution for the special case of $v_{max} = 1$ and are good approximations for higher values in the case of $v_{max} > 1$. In this paper, we use mean-field method [9] to analyze the NSOS model. The general express equations have been obtained in the stationary state in the case of $v_{max} > 2$. In the special case of $v_{max} = 1$, where overtaking vehicles can not overtake the preceding ones, the detail results in the stationary state have been done. In order to find the qualitatively different behaviors in the case of $v_{max} \geq 2$, we calculate the results for $v_{max} = 2$ where overtaking happens.

Figure 1. (Color online) Schematic graph of updating and moving in the NSOS model. The numbers in the sites represent the velocities after moving. The red cars are overtaking vehicles (left). The second car is an overtaking vehicle, it has tried to overtake the preceding one but failed (right). The first car is an overtaking vehicle, it has successfully overtook the preceding one.

The paper is organized as follows: First briefly introduce the update rules of the NSOS model in Section 2, then analyze this model by applying mean-field method in Section 3, and the summary is discussed in the final Section.

2. NSOS model

NSOS model is based on the NS model which considering overtaking strategy with probability q [1, 2]. In this model, overtaking vehicles are picked up randomly with probability q at every time, and they overtake preceding ones depending on their configurations in the next time. Some assumptions have been considered as followings: (1) Overtaking vehicles would brake if they reached the same location with their preceding ones in the next time to avoid collisions. (2) Each overtaking vehicle is only able to overtake one per time step. (3) The overtaking vehicle locates in front of preceding one once overtake successfully. Similar to NS model, all the overtaking vehicles decrease by one with braking probability p except for the successful overtaking ones. For convenience, we name the vehicles which are not overtaking as ordinary ones, and update their velocities according to the NS model. The detailed updating rules of the NSOS model are as follows:

1. At time t, the jth vehicle becomes an overtaking vehicle with probability q, otherwise it is an ordinary one.
2. Update the velocity:

(I) If the jth vehicle is an ordinary one:
(1) Acceleration:
$v(j, t_1) \to min(v(j, t) + 1, v_{max})$.
(2) Deceleration:
$v(j, t_2) \to min(v(j, t_1), d(j, t))$.
(3) Random braking:
$v(j, t_3) \to max(v(j, t_2) - 1, 0)$ with the probability p.
(II) If the jth vehicle is an overtaking one:
(1) Acceleration:
$v(j, t_1) \to min(v(j, t) + 1, v_{max})$.
(2) If $v(j, t_1) > d(j, t) + v(j+1, t+1)$, the position $d(j, t) + v(j+1, t+1) + 1$ is empty and the $(j + 1)$th vehicle does not overtake successfully,
(i) Overtaking:
$v(j, t_3) \to d(j, t) + v(j + 1, t + 1) + 1$.
(3) Otherwise,
(i) Deceleration:
$v(j, t_2) \to min(d(j, t) + v(j + 1, t + 1) - a, v(j, t_1))$.
(ii) Random braking with probability p:
$v(j, t_3) \to min(v(j, t_2) - 1, 0)$.

3. Movement:
$x(j, t + 1) = x(j, t) + v(j, t_3)$.

Here, $v(j, t)$ denotes the velocity of the jth vehicle at time t and $x(j, t)$ denotes its corresponding position. The number of empty sites in front of the jth vehicle is denoted by $d(j, t) = x(j + 1, t) - x(j, t) - 1$. To avoid collisions, we assume $a = 2$ if the $(j + 1)$th overtaking vehicle overtakes successfully. In other cases, $a = 1$. An illustration of the NSOS model can be found in Fig. 1. The numbers in the sites stand for their velocities after moving. In the left graph, the second car is an overtaking vehicle, it has tried to overtake the preceding one but failed. In the right graph, the first car is an overtaking vehicle, which has successfully overtook the preceding one.

Since the velocity of overtaking vehicle at time $t+1$ is relative to its preceding one, we should know the preceding vehicle's velocity at time $t + 1$ first. Fortunately, the velocity of ordinary vehicle is independent of the preceding one, so we could pick up ordinary vehicles and update their velocity first, and then the rear vehicles. In our simulations, we use parallel update and periodic boundary conditions, assume the first and last vehicles as ordinary ones all the time, update their velocities first, and then update others' velocities from $(N - 1)th$ car to the second one.

3. Mean-field theorem

The simplest analytical approach to the NS model is a microscopic mean-field (MF) theory [9]. Here one considers the probability $c_\alpha(i, t)$ of vehicles with velocity α at site i and time t, and we denote the probability that there is no vehicle at site i ($i = 1, 2, 3, ..., L$) at time t by $d(i, t)$. In the MF approach, correlations between sites are completely neglected. Therefore one has the normalization condition for all sites and all time steps,

$$d(i, t) + \sum_{\alpha=0}^{v_{max}} c_\alpha(i, t) = 1. \tag{1}$$

Denoting with $c(i, t)$ the total probability for site i to be occupied at time step t, i.e. $\sum_{\alpha=0}^{v_{max}} c_\alpha(i, t)$, one simply has $d(i, t) + c(i, t) = 1$. In our model each car could be an ordinary one with probability $\bar{q}$ (here, $\bar{q} = 1 - q$) or overtaking one with probability q. Therefore we have the equation $c = c^* + c^\dagger$, where $c^* = qc$ denotes the probability of being overtaking vehicles and $c^\dagger = \bar{q}c$ denotes the probability of being ordinary vehicles. Since the update rules of overtaking vehicles relate to the next time positions of preceding ones, the configurations of them should be obtained first. Here we adopt $C(i+j,t)$ ($D(i+j,t) = 1 - C(i+j,t)$) as the probability that there is (not) a vehicle at site $i+j$ ($j = 1, 2, 3, ..., v_{max}$) at the next time before one updates the site i at time t, need to say that, here, we just consider the vehicles in front of site i. We update the ordinary vehicles first, and choose the site of an ordinary vehicle as a starting site to update the rear vehicles.

According to the update rules, the time evolution of these probability distributions can be described by the following sets of equations:

(1) For the ordinary vehicles, the update rules are the same with the ones in the original NS model, therefore the MF equations for the stationary state ($t \to \infty$) read [9] :

$$\begin{aligned} c_0^\dagger &= (c + pd)c_0^\dagger + (1 + pd)c \sum_{\beta=1}^{v_{max}} c_\beta^\dagger, \\ c_\alpha^\dagger &= d^\alpha[\bar{p}c_{\alpha-1}^\dagger + (\bar{p}c + pd)c_\alpha^\dagger + (\bar{p} + pd)c \sum_{\beta=\alpha+1}^{v_{max}} c_\beta^\dagger], \qquad 0 < \alpha < v_{max} \\ c_{v_{max}}^\dagger &= \bar{p}d^{v_{max}}(c_{v_{max}-1}^\dagger + c_{v_{max}}^\dagger). \end{aligned} \tag{2}$$

(2) For the overtaking vehicles ($v_{max} > 2$):

(i) The acceleration stage

$$\begin{aligned} c_0^*(i,t_1) &= 0, \\ c_\alpha^*(i,t_1) &= c_{\alpha-1}^*(i,t), \qquad 0 < \alpha < v_{max} \\ c_{v_{max}}^*(i,t_1) &= c_{v_{max}}^*(i,t) + c_{v_{max}-1}^*(i,t). \end{aligned} \tag{3}$$

(ii) The deceleration stage:

$$\begin{aligned} c_0^*(i,t_2) &= c_0^*(i,t_1), \\ c_1^*(i,t_2) &= D(i+1,t)c_1^*(i,t_1) + D(i+1,t)C(i+2,t)c_2^*(i,t_1), \\ c_\alpha^*(i,t_2) &= \underbrace{c_\alpha^*(i,t_1)\prod_{j=1}^{\alpha} D(i+j,t)}_{\text{I}} + \underbrace{c_{\alpha+1}^*(i,t_1)\prod_{j=1}^{\alpha} D(i+j,t)C(i+\alpha+1,t)}_{\text{II}} \\ &\quad \underbrace{+ C(i+\alpha+2,t)C(i+\alpha+1,t)\prod_{j=1}^{\alpha} D(i+j,t)\sum_{\beta=\alpha+2}^{v_{max}} c_\beta^*(i,t_1)}_{\text{III}} \\ &\quad \underbrace{+ C(i+\alpha-1,t)D(i+\alpha,t)\prod_{j=1}^{\alpha-2} D(i+j,t)\sum_{\beta=\alpha}^{v_{max}} c_\beta^*(i,t_1)}_{\text{IV}}, \qquad 1 < \alpha < v_{max} \\ c_{v_{max}}^*(i,t_2) &= c_{v_{max}}^*(i,t_1)\prod_{j=1}^{v_{max}} D(i+j,t) + C(i+v_{max}-1,t) \\ &\quad \times D(i+v_{max},t)\prod_{j=1}^{v_{max}-2} D(i+j,t)c_{v_{max}}^*(i,t_1). \end{aligned} \tag{4}$$

(iii) The braking stage:

$$
\begin{aligned}
c_0^*(i,t_3) &= c_0^*(i,t_2) + p c_1^*(i,t_2),\\
c_1^*(i,t_3) &= \bar{p} c_1^*(i,t_2) + p[c_2^*(i,t_2) - C(i+1,t)D(i+2,t)\sum_{\beta=2}^{v_{max}} c_\beta^*(i,t_1)],\\
c_\alpha^*(i,t_3) &= \underbrace{p[c_{\alpha+1}^*(i,t_2) - C(i+\alpha,t)D(i+\alpha+1,t)\prod_{j=1}^{\alpha-1} D(i+j,t)\sum_{\beta=\alpha+1}^{v_{max}} c_\beta^*(i,t_1)]}_{\text{I}}\\
&\quad \underbrace{+\bar{p}[c_\alpha^*(i,t_2) - C(i+\alpha-1,t)D(i+\alpha,t)\prod_{j=1}^{\alpha-2} D(i+j,t)\sum_{\beta=\alpha}^{v_{max}} c_\beta^*(i,t_1)]}_{\text{II}}\\
&\quad \underbrace{+C(i+\alpha-1,t)D(i+\alpha,t)\prod_{j=1}^{\alpha-2} D(i+j,t)\sum_{\beta=\alpha}^{v_{max}} c_\beta^*(i,t_1),}_{\text{III}} \qquad 1<\alpha<v_{max}\\
c_{v_{max}}^*(i,t_3) &= \bar{p}[c_{v_{max}}^*(i,t_2) - C(i+v_{max}-1,t)D(i+v_{max},t)\prod_{j=1}^{v_{max}-2} D(i+j,t)c_{v_{max}}^*(i,t_1)]\\
&\quad + C(i+v_{max}-1,t)D(i+v_{max},t)\prod_{j=1}^{v_{max}-2} D(i+j,t)c_{v_{max}}^*(i,t_1).
\end{aligned} \tag{5}
$$

(iv) The motion stage:

$$c_\alpha^*(i,t+1) = c_\alpha^*(i-\alpha,t_3). \qquad 0 \le \alpha \le v_{max} \tag{6}$$

Since the overtaking vehicle is able to overtake the preceding one, it has more configurations than an ordinary vehicle. In the deceleration stage (ii), the item I describes that the overtaking vehicle at site i with velocity α will keep its velocity if the preceding one do not locate in these α sites in the next time. If the preceding vehicle will locate at the site $i+\alpha+1$ in the next time, the vehicle with velocity $\alpha+1$ decreases one to avoid collision (item II). The NSOS model forbids to overtake more than one, so stage (ii) contains an item III. If an overtaking vehicle overtakes its preceding vehicle, it will close to the overtaken one according to our model, this contributes a situation with velocity α when a vehicle overtakes one with velocity $\alpha-1$ expressed by item IV. In the braking stage (iii), all the overtaking vehicles decrease by one with braking probability p expect for successful overtaking vehicles, so one obtains an item I which the velocity with $\alpha+1$ of an overtaking vehicle minuses one with probability p except for successful overtaking ones. Similarly, item II can be obtained without braking and the successful overtaking vehicle with velocity α also has a contribution to this stage (item III).

Even though these time evolution equations are nonlinear, in the limit $t \to \infty$, the C and D distributions become homogeneous in space (for periodic boundary conditions). Therefore we can use the C's values apart from the time and site dependences to calculate the traffic flow.

$$
\begin{aligned}
c_0^* &= pDc_0^* + pDCc_1^* + pDC^2 \sum_{\beta=2}^{v_{max}} c_\beta^*, \\
c_1^* &= \bar{p}Dc_0^* + (\bar{p}DC + pD^2)c_1^* + pD^2Cc_2^* + pD^2C^2 \sum_{\beta=3}^{v_{max}} c_\beta^*, \\
c_\alpha^* &= D^{\alpha-1}[(\bar{p}D + C)c_{\alpha-1}^* + (\bar{p}DC + pD^2 + C)c_\alpha^* \\
&\quad + (\bar{p}DC + pD^2 + 1)Cc_{\alpha+1}^* + (\bar{p}DC + pD^2C + 1)C \sum_{\beta=\alpha+2}^{v_{max}} c_\beta^*], \qquad 1 < \alpha < v_{max} \\
c_{v_{max}}^* &= D^{v_{max}-1}(\bar{p}D + C)(c_{v_{max}}^* + c_{v_{max}-1}^*).
\end{aligned} \tag{7}
$$

These equations are linear when we apply the relation $C = 1 - D$. So the equations (2) and (7) can be recast in matrix form as $\mathbf{M}\vec{c} = \vec{c}$. The matrix $\mathbf{M}$ can be read off from (2) and (7), $\vec{c}$ is the vector with elements c_α, $\alpha = 0, \ldots, v_{max}$. For small v_{max} one can calculate the probability c_α explicitly.

Since $c^\dagger = \bar{q}c$, $c^* = qc$ and $c = c^\dagger + c^*$, using the equations (2) and (7), we could obtain the specific form of c_α $(0 \leq \alpha \leq v_{max})$. Moreover, combined with the equation of flow $f(c, p, q) = \sum_{\alpha=1}^{v_{max}} \alpha c_\alpha$, we can calculate the flow as a function of p, q and c. Next, we calculate the flow in the case of $v_{max} = 1$ and $v_{max} = 2$, respectively.

3.1. *Takeover:* $v_{max} = 1$

No overtaking vehicles could overtake the preceding one in the limit of $v_{max} = 1$. In fact, it becomes the takeover case which is discussed in the paper [14], that is to say, overtaking vehicles could advance to the position that was occupied by their preceding ones at the previous time step.

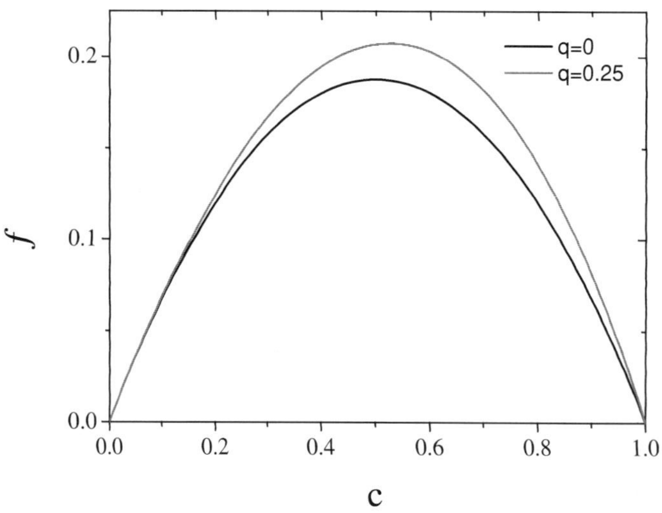

Figure 2. (Color online) Fundamental diagram flow f vs density c for maximum velocity $v_{max} = 1$ in the mean-field approximation in the case of $p = 0.25$. Black line is in the case of $q = 0$, and red line is in the case of $q = 0.25$.

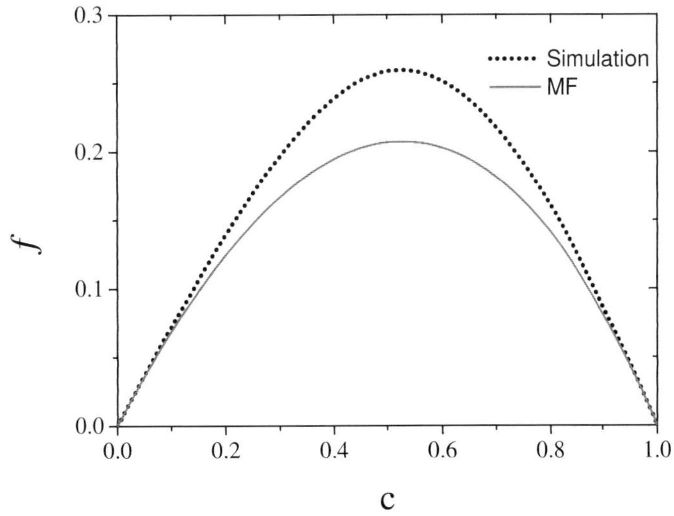

Figure 3. (Color online) Fundamental diagram for $v_{max} = 1$, $p = 0.25$ and $q = 0.25$. The red full curve is the MF result. For comparison the result from computer simulation (black dots) is also shown.

According to the update steps of the NSOS model, the time evolutions of these probability distributions can be described by the following four sets of equations:

(1) For the ordinary vehicles with $v_{max} = 1$, the MF equations for the stationary state $(t \to \infty)$ are given by [9]

$$\begin{aligned} c_0^\dagger &= (c + pd)c^\dagger, \\ c_1^\dagger &= \bar{p}dc^\dagger. \end{aligned} \tag{8}$$

with $c^\dagger = c_0^\dagger + c_1^\dagger$.

(2) For the overtaking vehicles:

$$\begin{aligned} c_0^* &= pDc^*, \\ c_1^* &= \bar{p}Dc^*. \end{aligned} \tag{9}$$

Since $c^\dagger = \bar{q}c$, $c^* = pc$ and $c = c^\dagger + c^*$, we have

$$\begin{aligned} c_0 &= (c + pd)\bar{q}c + pqDc, \\ c_1 &= \bar{p}\bar{q}dc + \bar{p}qDc. \end{aligned} \tag{10}$$

The flow $f(c, p, q)$ in the case $v_{max} = 1$ is

$$f(c, p, q) = c_1 = \bar{p}\bar{q}dc + \bar{p}qDc. \tag{11}$$

According to the definition of D, which denotes the empty possibility in the next time, considered the mean-field approximation it equals to the sum of distance d of two neighborhood vehicles with the flow f, i.e. $D \approx d + f$. So the equation becomes

$$f(c, p, q) = \frac{\bar{p}dc}{1 - \bar{p}qc}. \tag{12}$$

The first information from this equation is that the flow is dominated by braking probability p, while the overtaking probability q is not the important factor since q has a factor $\bar{p}c$. Specially, the flow becomes the form of NS model when $q = 0$. Another finding is that the flow is enhanced mainly in the jammed phase. If we keep the p and q invariant, with the growth of density c, the denominator decreases monotonically, while the value of flow increases larger. This result can be explained in physical terms. In the free flow phase, all the vehicles move freely which means the preceding ones has no influence to overtaking vehicles, the impact of D on the flow is the same as the one of d. While in the jammed phase, overtaking vehicles have larger probabilities to move than ordinary ones due to $D > d$, so the flow in this phase is larger than the one of NS model. Moreover, the larger q induces, the quicker flow increases.

The mean-field result yields, compared with the simulation data shown in Fig. 3, much too small values of the flow. This can easily be understood since the reduction to a single vehicles problem ignores all spatial correlations of the vehicles [9].

3.2. Overtaking: $v_{max} = 2$

It is the simplest case that overtaking vehicles could overtake the preceding ones in the NSOS model in the case of $v_{max} = 2$. Moreover, it is natural without innateness hypothesis that an overtaking vehicle is only able to overtake one each time and occupy the site just in front of its preceding one if it could overtake successfully. So it is necessary to calculate the exact equations in the case of $v_{max} = 2$. Using the (2)-(5) equations, we have:

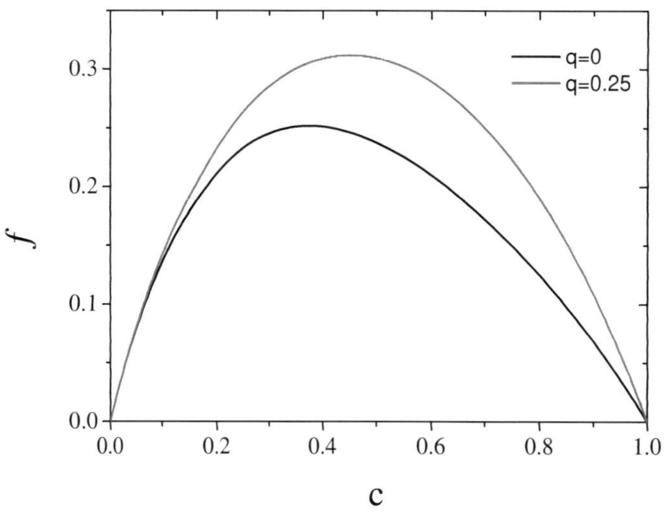

Figure 4. (Color online) Fundamental diagram flow f vs. density c for maximum velocity $v_{max} = 2$ in the mean-field approximation in the case of $p = 0.25$. Black line is in the case of $q = 0$, and red line is in the case of $q = 0.25$.

(1) For the ordinary vehicles:

$$\begin{aligned} c_0^\dagger &= \frac{(1+pd)c}{1-pd^2}c^\dagger, \\ c_1^\dagger &= \frac{\bar{p}(1-\bar{p}d^2)d}{1-pd^2}c^\dagger, \\ c_2^\dagger &= \frac{\bar{p}^2 d^3}{1-pd^2}c^\dagger. \end{aligned} \tag{13}$$

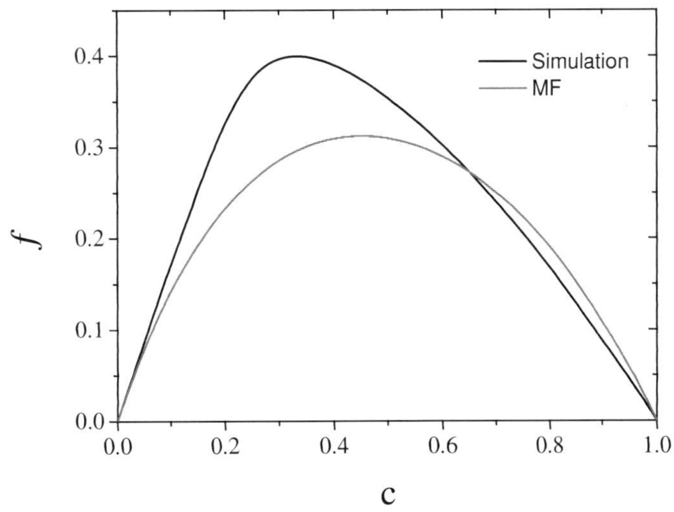

Figure 5. (Color online) Fundamental diagram for $v_{max} = 2$, $p = 0.25$ and $q = 0.25$. The red full curve is the MF result. For comparison the result from computer simulation (black dots) is also shown.

(2) For the overtaking vehicles:

$$\begin{aligned} c_0^* &= \frac{pDC}{1-pD^2}c^*, \\ c_1^* &= \frac{\bar{p}C + pD(1-pD)}{1-pD^2}Dc^*, \\ c_2^* &= \frac{(\bar{p}D + C)(1-pD)}{1-pD^2}Dc^*. \end{aligned} \tag{14}$$

Here, we have used equations $C + D = 1$ and $c_1^* + c_2^* = c^* - c_0^*$. Again, we could calculate c_α ($\alpha = 0, 1, 2$) using $c_\alpha = c_\alpha^\dagger + c_\alpha^*$ and $c^\dagger = \bar{q}c$, $c^* = qc$. But this time we assume $D \approx d$ due to the value of $D - d$ is smaller than d. The flow can be calculated using the following equation $f(c, p, q) = (c_1^\dagger + c_1^*) + 2(c_2^\dagger + c_2^*)$. The result is shown in Fig. 4. One could also observe that the flow of the NSOS model enlarged in the jammed regime than that of original NS model, this would be due to overtaking mechanism is beneficial to develop the traffic flow. Again, the mean-filed result is still less than simulation data (Fig. 5), and one could observe that the maximum flow density does not coincide with the simulation result, this may be the result of our simplicity D.

4. Conclusions

In this paper theoretical analysis of the NSOS model is performed by using the mean-field method, the equations for $v_{max} = 1$ can be obtain exactly, while for larger values of v_{max} they are just approximations. Even though mean-field theory is insufficient due to the important correlations between neighboring sites are neglected, the reason that why the NSOS model can improve traffic flow in the area where the flow exceed the maximum flow density has been explained, and braking probability as a major factor that influence transition density has been discovered.

5. Acknowledgments
This work was supported by China Postdoctoral Science Foundation (Grant Nos. 30205010003), self-determined research funds of CCNU from the colleges' basic research and operation of MOE (Grant Nos. KJ02072017-2291), and National Natural Science Foundation of China (Grant Nos. 11505071 and 61702207).

Appendix A: Takeover Case
In this appendix we show that the stationary state of the NSOS model with $v_{max} = 1$ for overtaking vehicles:

(i) The acceleration stage:

$$\begin{aligned} c_0^*(i,t_1) &= 0, \\ c_1^*(i,t_1) &= c_1^*(i,t) + c_0^*(i,t). \end{aligned} \tag{15}$$

(ii) The deceleration stage:

$$\begin{aligned} c_0^*(i,t_2) &= c_0^*(i,t_1), \\ c_1^*(i,t_2) &= D(i+1,t)c_1^*(i,t_1). \end{aligned} \tag{16}$$

(iii) The braking stage:

$$\begin{aligned} c_0^*(i,t_3) &= c_0^*(i,t_2) + pc_1^*(i,t_2), \\ c_1^*(i,t_3) &= \bar{p}c_1^*(i,t_2). \end{aligned} \tag{17}$$

(iv) The motion stage:

$$c_\alpha^*(i,t+1) = c_\alpha^*(i-\alpha,t_3), \qquad 0 \le \alpha \le 1. \tag{18}$$

In the stationary state, distributions become homogeneous in space for periodic boundary conditions, so the site dependence could be omitted. Using this and combining the four update steps one gets the set of equations (10).

Appendix B: Overtaking Case
In this appendix we show that the stationary state of the NSOS model with $v_{max} = 2$ for overtaking vehicles:

(i) The acceleration stage:

$$\begin{aligned} c_0^*(i,t_1) &= 0, \\ c_1^*(i,t_1) &= c_0^*(i,t), \\ c_2^*(i,t_1) &= c_2^*(i,t) + c_1^*(i,t). \end{aligned} \tag{19}$$

(ii) The deceleration stage:

$$\begin{aligned} c_0^*(i,t_2) &= c_0^*(i,t_1), \\ c_1^*(i,t_2) &= D(i+1,t)c_1^*(i,t_1) + D(i+1,t)C(i+2,t)c_2^*(i,t_1), \\ c_2^*(i,t_2) &= D(i+1,t)D(i+2,t)c_2^*(i,t_2) + C(i+1,t)D(i+2,t)c_2^*(i,t_1). \end{aligned} \tag{20}$$

(iii) The braking stage:

$$\begin{aligned}
c_0^*(i,t_3) &= c_0^*(i,t_2) + pc_1^*(i,t_2), \\
c_1^*(i,t_3) &= \bar{p}c_1^*(i,t_2) + p[c_2^*(i,t_2) - C(i+1,t)D(i+2,t)c_2^*(i,t_1)], \\
c_2^*(i,t_3) &= \bar{p}[c_2^*(i,t_2) - C(i+1,t)D(i+2,t)c_2^*(i,t_1)] + C(i+1,t)D(i+2,t)c_2^*(i,t_1).
\end{aligned} \tag{21}$$

(iv) The motion stage:

$$c_\alpha^*(i,t+1) = c_\alpha^*(i-\alpha,t_3), \qquad 0 \le \alpha \le 1. \tag{22}$$

In the stationary state, distributions become homogeneous in space for periodic boundary conditions, so the site dependence could be omitted. Using this and combining the four update steps one gets the set of equations (14).

References

[1] Zhu S, Weibing D, Jihui H, Wei L and Xu C 2016 *International Journal of Modern Physics C* **27** 1-12.
[2] Zhu S, Weibing D, Longfeng Z, Jihui H, Wei L and Xu C 2016 *European Physical Journal B* **89** 203.
[3] Debashish C, Ludger S and Schadschneider A 2000 *Phys. Rep.* **329** 199-329.
[4] Helbing D 2001 *Rev. Mod. Phys.* **73** 1067-1141.
[5] Wolfram S 1983 *Rev. Mod. Phys.* **55** 601-644.
[6] Nagel K and Schreckenberg M 1992 *J. Phys. I France* **2** 2221-2229.
[7] Schadschneider A and Schreckenberg M 1997 *J. Phys. A: Math. Gen.* **10** L69-L75.
[8] Schadschneider A 1999 *Eur. Phys. J. B* **10** 573-582.
[9] Schreckenberg M, Schadschneider A, Nagel K and Ito N 1995 *Phys. Rev. E* **51** 2939-2949.
[10] Schadschneider A and Schreckenberg M 1993 *Journal of Physics A: Mathematical and General* **26** L679.
[11] Schadschneider A and Schreckenberg M 1997 *Journal of Physics A: Mathematical and General* **30** L69.
[12] Schadschneider A and Schreckenberg M 1998 *Journal of Physics A: Mathematical and General* **31** L225.
[13] Tilstra L G and Ernst M H 1998 *Journal of Physics A: Mathematical and General* **31** 5033.
[14] HUNG JUNG C 2001 *Int. J. Mod. Phys. B* **15** 3453-3464.

The stock market learned as Ising model

Longfeng Zhao[1], Weiqi Bao[2], Wei Li[1]

[1] Key Laboratory of Quark and Lepton Physics (MOE) and Institute of Particle Physics, Central China Normal University, Wuhan 430079, China

[2] Commercial College, Shandong University, Weihai, Weihai 264209, China

E-mail: zlfccnu@mails.ccnu.edu.cn (L. Zhao), liw@mail.ccnu.edu.cn (W. Li)

Abstract. The collective behaviors of the financial systems resemble the physics theory of critical phenomena. Here we treat the stock market as the famous phase transition model: Ising model and the model parameters are learned from the real stock return time series. A comparative analysis between three world major stock markets have been given.

1. Introduction

The research about the collective behaviors of the complex financial system are extremely attractive. There are diverse methods been put forward to explore the complexity of financial systems [1]. The applications of those methods from statistical physics have provided us many new insights about financial systems [2–7]. The analogy between financial system, e.g. stock market, and real physical system such as spin system is a very interesting topic not only for physicists but also for financial industry.

The stock price are formed through the exchanges among many market participants. Thus the change in prices are correlated due to endogenous and exogenous events shared by many stocks. The inverse Ising problems offered us very flexible tools to infer the interaction patterns from the empirical data [8]. In ref. [9], three different inference methods have been used to obtain the interaction strength of stocks from the traded volumes. The authors show that the traded volumes are informative enough for the inference of collective behaviors. In refs. [10–12], Bury mapped the stock interactions to a spin model with pairwise interactions. Some cluster structure based on the coupling matrix have been uncovered with very clear economic meaning. Although extensive studies have been proceeded with the presence of large amount of financial data [9–14]. A comparative analysis between different countries is still lacking.

Here we try to infer the interaction strength of the stock markets from empirical data by recognizing the market as a spin glass model. We use the stock return time series from three markets for different countries. The statistical properties of the coupling matrix have been analyzed from which some common patterns have been revealed.

The paper is organized as follows. In Sec. 2 we introduce the methodologies used in this paper. In Sec. 3 we present the main empirical results. The last section provides our conclusion and discussion.

2. Data and Methodology

2.1. Data

Our datasets are consist of the daily adjusted returns time series of the constitute stocks from three major stock indexes in the world: S&P 500 (the US), FTSE 350 (the UK) and SSE 380 (China). After removing those stocks with relative small sample sizes, we still have 401, 264, and 295 stocks for the three markets respectively. We choose those indexes to make the system size as large as possible. In the S&P 500 dataset, each stock has 4025 daily records from 4 January 1999 to 31 December 2014. The FTSE 350 stocks include 3000 daily returns in the period between 10 October 2005 and 26 April 2017. The SSE 380 stocks consist of 2700 daily returns from 21 May 2004 to 19 November 2014.

In this paper, we use the logarithm return $r_i(t)$ which is defined as

$$r_i(t) = \ln(p_i(t+1)) - \ln(p_i(t)). \tag{1}$$

$p_i(t)$ is the adjusted closure price of stock i at time t. Then the return sequence are binarized as the state of each stock or spin:

$$s_i(t) = 1, r_i(t) \geq 0; s_i(t) = -1, r_i(t) < 0. \tag{2}$$

Here $s_i(t)$ is the state of stock i at time t. In our empirical analysis, $r_i(t) = 0$ is very rare. We then use the state sequence of those stocks to infer the Ising model interaction strengths and external fields.

2.2. Definition of the problem

As we mentioned in the previous context, we make an analogy between the financial system and Ising model. The binarized states of each stocks in three stock markets have been directly mapped to a two state Ising spin model from which the two point cross-correlation should be reconstructed during the inference process.

We consider the Ising model with N binary spin variables $\{s_i = \pm 1\}, i = 1, \ldots, N$. The pairwise couplings J_{ij} determine the interactions between spins. The Hamiltonian of the system with a spin configuration $\mathbf{s} = \{s_i\}$ is specified by

$$H_{\mathbf{J},\mathbf{h}}(\mathbf{s}) = -\sum_{i<j} J_{ij} s_i s_j - \sum_i h_i s_i. \tag{3}$$

We then introduce two methods borrowing from inverse problem researches which can help us reconstruct the interaction strength and external fields of Ising model.

2.3. naive mean field

The exact learning is general slow and can be substituted by the approximate inferences methods which are based on expansion of the free energy of a system for small fluctuations around its mean value. The equation of the coupling strength J can be derived form the second order derivation of free energy [8]:

$$\mathbf{J}^{nMF} = \mathbf{A}^{-1} - \mathbf{C}^{-1}. \tag{4}$$

Similarity, we can also get the external field of the system which is written as follows:

$$h_i^{nMF} = \text{artanh}\langle s_i \rangle - \sum_{j=1}^{N} J_{ij}^{nMF} \langle s_i \rangle. \tag{5}$$

Here $\mathbf{A}$ is a diag matrix with elements $A_{ii} = 1 - \langle s_i \rangle^2$ and $\mathbf{C}$ is the covariance of the state sequences of the system. This is called the naive mean field method(nMF).

2.4. TAP approximation

The second-order correction to the nMF approximation requires solving Thouless-Anderson-Palmer (TAP) equations. In 1977, Thouless, Anderson, and Palmer (TAP) added a term to the Gibbs free energy [8]. Thus we have

$$(\mathbf{C}^{-1})_{ij} = -J_{ij}^{TAP} - 2(J_{ij}^{TAP})^2\langle s_i\rangle\langle s_j\rangle, \tag{6}$$

$$h_i^{TAP} = h_i^{nMF} - \langle s_i\rangle \sum_{j=1}^{N} (J_{ij}^{TAP})^2(1-\langle s_j\rangle^2). \tag{7}$$

Then we can get the coupling strength of the system as follows:

$$J_{ij}^{TAP} = \frac{-2(\mathbf{C}^{-1})_{ij}}{1+\sqrt{1-8(\mathbf{C}^{-1})_{ij}s_is_j}}. \tag{8}$$

From Eq(4) and (8), we know the the coupling strength depends on the inverse of covariance matrix **C**. The only way to guarantee the invertibility of a covariance matrix is to make the length of the time series as long as possible. Here the lengths of the time series are significantly larger than the system size. Thus the covariance **C** is invertible.

3. Results

Here in Fig. 1 we have calculated the distributions of the coupling strength **J** for three markets. The black lines are the coupling strength distributions of the three markets and for comparison purpose, we also simulate the shuffled return time series and the red lines denoted the coupling strength distributions of three markets for the shuffled case. The shuffling procedure is proceed by randomly exchange the stock return time series. Thus we want to remove the effect of cross-correlations related to the collective behaviors. It's obvious that the coupling strength of three markets largely deviated from the shuffled case, which can not be recognized as Edwards-Anderson model with Gaussian distributed coupling strengths. The heterogeneity coupling strengths give an hint about the criticality properties of the stock markets which is common for three markets.

Table 1: The first four statistical moments for the distributions of interaction strength J_{ij} for both mean field and TAP methods. The first value in each cell is the statistical moments of J_{ij} for mean field method and the second value for TAP method.

	mean	variance	skewness	kurtosis
SP 500	0.0045;0.0045	0.0015;0.0015	1.95;1.94	23.19;23.16
FTSE 350	0.0052;0.0052	0.0016;0.0016	2.62;2.62	32.93;32.83
SSE 380	0.0054;0.0054	0.0020;0.0020	1.07;1.06	8.86;8.83

In table1, we present the first four order statistical moments for the distribution of the interaction strength J_{ij}. The skewness of SP 500 and FTSE 350 markets are twice as much as the value of SSE 380. This means the interaction strengths of SP 500 and FTSE 380 are right skewed compare with SSE 380. The kurtosis of SP 500 and FTSE 350 are extremely larger than that of SSE 380. Thus we know the interaction strength distribution of SP 500 and FTSE 350 are very heterogeneous and heavy tailed.

Fig.2 shows the reconstructed external field h for three markets. The distributions have their own identities. The external field for US market has relative wider and flatten distribution. But

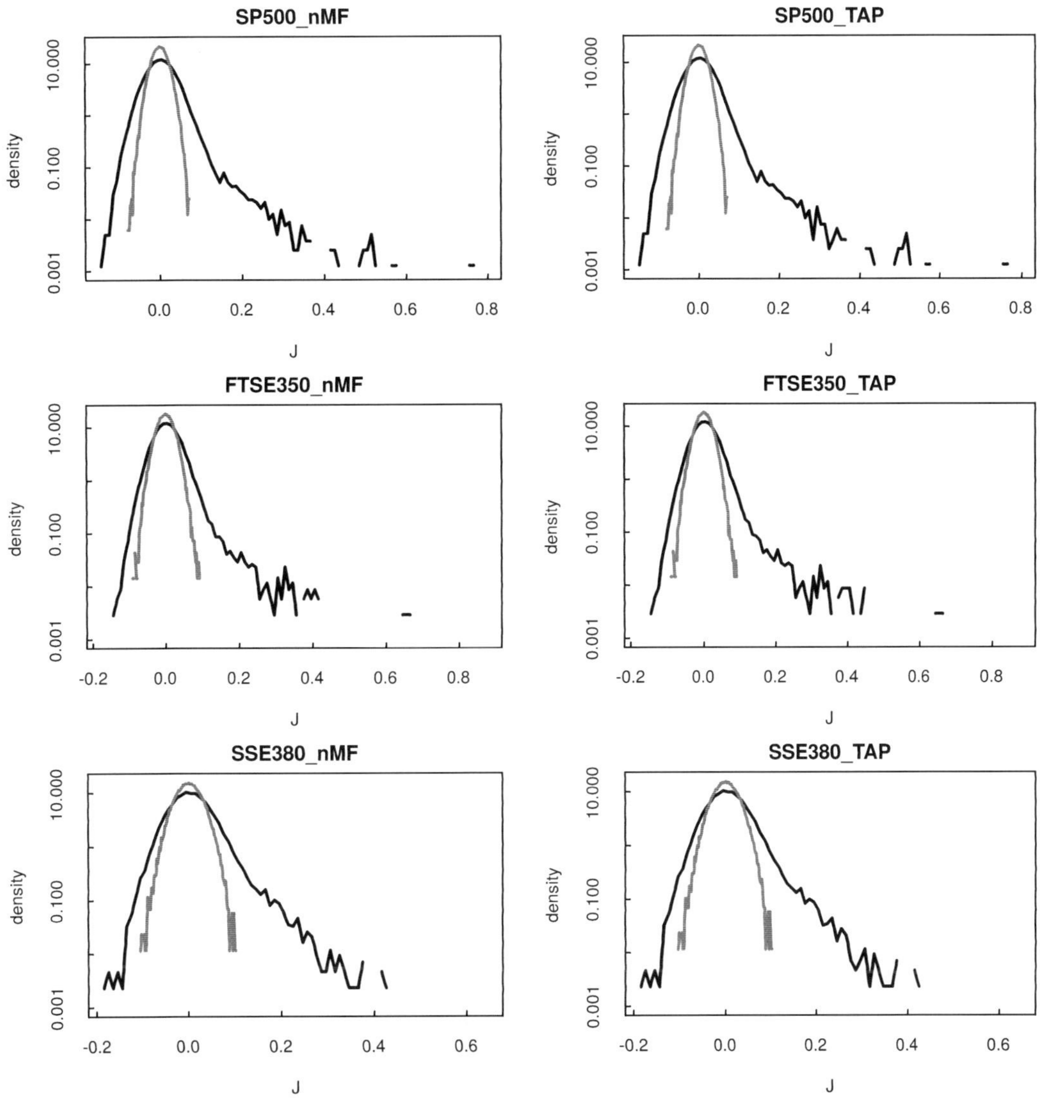

Figure 1: The interaction strength distributions of the inferred Ising spin system for three different markets. We use both mean field and TAP approximation methods. The left column is the results for three stock markets with mean field method and the right column is the result with TAP approximation method.

the external field for the UK market is more peaked. And the fluctuation of the distribution of China market is relatively larger than other two western markets. This maybe interpreted as the immaturity of China market.

Here in table2 we give the first four order statistical moments for the distributions of external field h_i for both mean field and TAP methods. From the statistical moments we know that the external fields recovery form empirical data are very close to white noise with very small mean, variance and skewness. Also the kurtosis are also very close to the value of Gaussian distribution with kurtosis 3.

The discussion about the market criticality is always a very attractive topic among

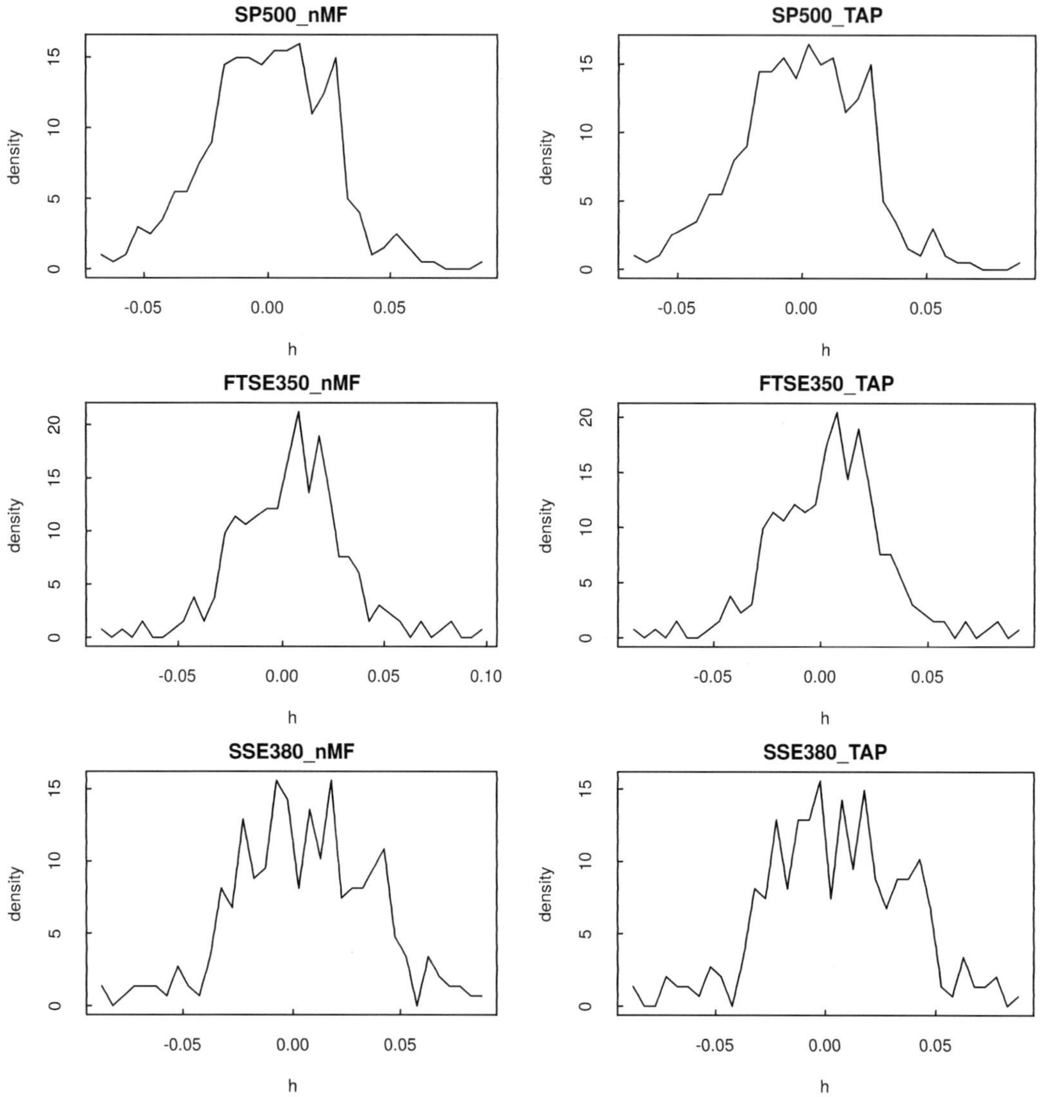

Figure 2: The external field distributions for three markets inferred with both mean field and TAP approximation methods. The left column is the results for three stock markets with mean field method and the right column is the result with TAP approximation method.

Table 2: The first four statistical moments for the distribution of external field h_i for both mean field and TAP methods. The first value in each cell is the statistical moments ofh_i for mean field method and the second value for TAP method.

	mean	variance	skewness	kurtosis
SP 500	0.0001;-0.00013	0.00061;0.00060	-0.0097;-0.0082	3.10;3.10
FTSE 350	0.0039;0.0037	0.00070;0.00069	0.053;0.055	4.18;4.17
SSE 380	0.0053;0.0047	0.0010;0.00098	-0.12;-0.11	3.09;3.10

econophysics community. Here the heterogeneous interaction strengths can be seen as a characteristic of criticality which definitely requires further investigation and validation from multiple perspectives. But the fingerprint provided here shows universal criticality of the world wide stock markets.

4. Conclusion

In conclusion, we use the binarized approach to transform the return time series of three major stock markets into Ising spin states with ± 1. The analogy between Ising model and stock market has revealed the non-Gaussian properties of the stock market interactions. The coupling strength distributions and the external field distributions have shown market specific fingerprints. The detailed comparison analysis among different stock markets from the view point of critical pairwise interaction model are very attractive and more attention about this topic is called for. A detailed comparison analysis based on Monte Carlo simulation will be given in the future research and the detailed analysis about the interaction matrix dynamics based on eigenvector dynamics and complex networks should also be interesting directions [3, 15].

Acknowledgement

This work is supported in part by the Programme of Introducing Talents of Discipline to Universities under grant NO. B08033 and the program of China Scholarship Council (No. 201606770023).

5. References

[1] Kwapień J and Drożdż S 2012 *Physics Reports* **515** 115–226
[2] Chakraborti A, Toke I M, Patriarca M and Abergel F 2011 *Quantitative Finance* **11** 991–1012
[3] Zhao L, Li W and Cai X 2016 *Physics Letters A* **380** 654–666
[4] Xin C, Yang G and Huang J P 2017 *Physica A: Statistical Mechanics and its Applications* **471** 666–673
[5] Tsallis C 2017 *Entropy* **19** 1–17
[6] Zhao L, Li W, Fenu A, Podobnik B, Wang Y and Stanley H E 2018 *Journal of Statistical Mechanics: Theory and Experiment* **2018** 23402
[7] Zhao L, Wang G J, Wang M, Bao W, Li W and Stanley H E 2018 *Physica A*, *https://doi.org/10.1016/j.physa.2018.05.039.*
[8] Nguyen H C, Zecchina R and Berg J 2017 *Advances in Physics* **66** 197–261
[9] Zeng H, Lemoy R and Alava M 2014 *Journal of Statistal Mechanics: Theory and Experiment* **07** P07008
[10] Bury T 2013 *European Physical Journal B* **86** 7
[11] Bury T 2013 *Journal of Statistical Mechanics: Theory and Experiment* **2013** P11004
[12] Bury T 2013 *Physica A: Statistical Mechanics and its Applications* **392** 1375–1385
[13] Mastromatteo I, Zarinelli E and Marsili M 2011 *Journal of Statistal Mechanics: Theory and Experiment*
[14] Mastromatteo I and Marsili M 2011 *Journal of Statistal Mechanics: Theory and Experiment*
[15] Marti G, Nielsen F, Bińkowski M and Donnat P 2017 *arXiv* 1–25

IOP Conf. Series: Journal of Physics: Conf. Series **1113** (2018) 012010 doi:10.1088/1742-6596/1113/1/012010

The robustness of worldwide metro networks

Longfeng Zhao[1], **Wei Li**[1], **Yueying Zhu**[1]

[1] Key Laboratory of Quark and Lepton Physics (MOE) and Institute of Particle Physics, Central China Normal University, Wuhan 430079,China

E-mail: zlfccnu@mails.ccnu.edu.cn (L. Zhao), liw@mail.ccnu.edu.cn (W. Li)

Abstract. The robustness of worldwide metro networks have been revealed by using random attack and target attack strategies. We find that the metro networks are more robust towards random attack. The robustness of those metro networks have also been compared with two model networks: random networks and scale-free networks. The metro networks are more robust than scale-free networks under target attack but more fragile under random attack. Comparing to random networks, the metro networks are vulnerable under both random and target attacks.

1. Introduction

Our daily life are networked by many infrastructure systems among which the most important one is the transportation network[1]. Metro network is the most important part of modern urban transportation. The transportation cost of metro network is relatively low compare with other commuter ways such as bus and self-driving. The low commute time of metro network relies on its stability. Thus its robustness is of great importance.

Recently many operation failures of metro networks have been reported such as the 2010 bombings in Moscow's subway network and the 2008 London subway accident. Those are malicious terrorist attacks. Thus it has been extremely important to quantify the robustness of the metro networks from both random failure and target attack.

The complex network theory has taken over most of the science ranging from physics, biology to economics from nearly two decades ago. Since then its wide application to those fields have been extremely fruitful[2, 3, 4, 5, 6]. Thus the use of network representation to characterize the metro networks is quite straightforward. In this paper, we have analyzed the robustness of 28 worldwide metro networks under random attack and target attack. The largest connected component is used as a metric of functionality. We found that the metro networks are relatively robust under random attack compare to target attack. We also simulate two different model networks: Barabási-Albert model(BA network) and Erdős-Rényi model(ER network) as reference models[7, 8]. When comparing to BA networks, the metro networks are more vulnerable under random attack and more robust under target attack. But when comparing to ER networks, the metro networks are more vulnerable under both random and target attacks. The paper is organized as follows. In Sec. 2 we present the basic statistics of the metro networks. In Sec. 3 we give the main empirical results. The last section provides our conclusion.

2. Basic statistics of metro networks

We have analyzed 28 worldwide metro networks. The basic statistics of those metro networks have been listed below. The basic topological structures of those metro networks have been

IOP Conf. Series: Journal of Physics: Conf. Series **1113** (2018) 012010 doi:10.1088/1742-6596/1113/1/012010

discussed in our previous paper[9] and the data source has also been provided therein.

Table 1. Topological parameters of those 28 metro networks. N is the number of stations for each metro network; N_{L} is the number of lines; m is the number of edges; $\langle k \rangle$ is the average degree; $\langle s \rangle$ is the average strength of the metro networks; ρ is the density of the metro networks; l is the average shortest-path length; C is the clustering coefficient; $N_{\triangle}$ and $N_{\square}$ are the number of 3-node loops and the number of 4-node loops.

Cities	N	N_{L}	m	$\langle k \rangle$	$\langle s \rangle$	ρ	l	C	$N_{\triangle}$	$N_{\square}$
Barcelona	137	12	153	2.234	2.277	0.016	10.871	0.037	3	4
Beijing	225	16	249	2.213	2.222	0.010	14.419	0	0	3
Berlin	307	24	356	2.319	3.225	0.008	12.681	0.015	3	6
Boston	120	11	127	2.117	2.217	0.018	12.858	0.033	2	4
Busan	120	5	123	2.050	2.050	0.017	14.962	0	0	0
Chicago	157	9	177	2.255	2.408	0.015	12.353	0.037	4	16
Delhi	137	8	138	2.015	2.015	0.015	16.088	0	0	0
Guangzhou	129	9	135	2.093	2.093	0.016	13.398	0	0	2
Hamburg	150	11	169	2.253	3.107	0.015	9.893	0.031	3	2
Hong Kong	84	10	87	2.071	2.214	0.025	11.064	0.027	1	0
London	289	21	342	2.367	2.547	0.008	13.550	0.041	9	21
Madrid	275	16	312	2.269	2.291	0.008	15.727	0.030	5	13
Melbourne	216	17	220	2.037	3.361	0.010	22.227	0.035	3	1
Mexico	163	12	183	2.245	2.245	0.014	11.728	0.011	1	2
Milan	210	17	230	2.190	2.724	0.011	10.885	0	0	2
Montreal	68	4	69	2.029	2.029	0.030	10.595	0	0	0
Moscow	150	13	174	2.320	2.333	0.016	9.916	0.066	7	11
New York	595	24	741	2.491	2.807	0.004	12.312	0.032	20	56
Osaka	97	8	112	2.309	2.371	0.024	8.335	0	0	4
Paris	301	18	357	2.372	2.445	0.008	12.079	0.031	7	12
Seoul	497	23	567	2.282	2.511	0.005	18.922	0.020	6	5
Shanghai	251	14	275	2.191	2.271	0.009	14.986	0.007	1	2
Shenzhen	118	5	126	2.136	2.136	0.018	13.618	0.018	1	1
Taipei	101	13	103	2.040	2.040	0.020	12.312	0	0	0
Tokyo	216	13	261	2.417	2.574	0.011	10.543	0.048	8	7
Valencia	141	10	146	2.071	2.596	0.015	15.317	0.017	1	0
Vienna	90	5	96	2.133	2.133	0.024	9.929	0	0	1
Washington	86	6	88	2.047	3.023	0.024	11.065	0	0	0

In table 1 the basic topological parameters have been given. The number of stations or the number of nodes N in the metro networks range from 68 for Montreal to 595 for New York. Although the number of lines in the metro networks N_{L} and number of links m vary from city to city, but the average degree $\langle k \rangle$ of those metro networks are very close to 2. The lack of 3-node loops $N_{\triangle}$ and 4-node loops $N_{\square}$ make those metro networks very similar to tree graphs with very small clustering coefficient C, relatively large shortest path length l and quite sparse network density ρ. Those topological quantities give us a very detailed description of the geometrical structures of the metro networks. In the following context we will try to evaluate the robustness of those metro networks from two attack strategies: random attack and target attack. Two model networks have also been simulated and used as reference models to quantify the robustness of metro networks.

3. Robustness of metro networks

Here we use two attack strategies to model the failure of metro networks. The first one is random attack strategy which means there are $1 - p$ fraction of nodes will be randomly removed from the metro networks. The second one is the target attack strategy. Here we use the degree based attack strategy which means $1-p$ fraction nodes will be removed based on their degree ranking. Thus after the attack, there are only p fraction of nodes left in the metro networks. We use the largest connect component G as a metric of robustness of the networks. In this way, we assume that only the largest connected component can remain functional after the attack.

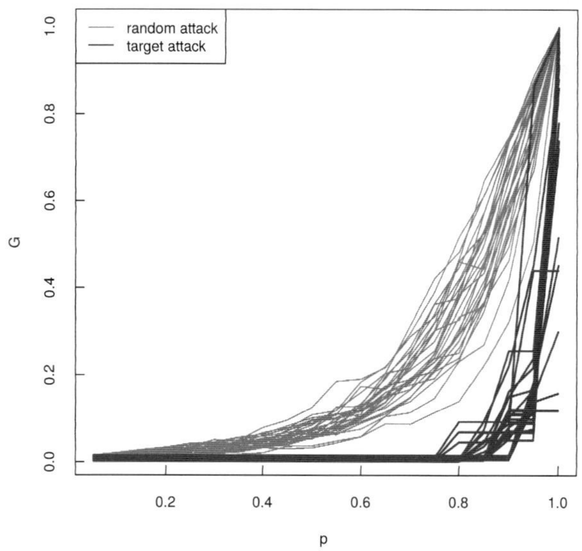

Figure 1. The largest connected components of 28 metro networks under both random attack and target attack. The red lines denote the largest connected components as a function of remaining node fraction p under random attack strategy and the blue lines represent the largest connected components under target attack strategy.

In figure 1, we have given the largest connected component G of those 28 metro networks under both random and target attacks. The red lines are the largest connected component G as a function of remaining node fraction p under random attack and the blue lines are the results for target attack. Very distinctive patterns for different attack strategies manifest the robustness of metro networks under different failure types. Obviously we see that the metro networks are more robust under random failure. The very sharp decrease of the largest connect component G under degree based attack gives a hint about the importance of those hub nodes in the metro networks. Those stations are extremely important for the maintainance of the functionality. Thus those large degree stations should be taken very strict protection.

In order to further quantify the robustness of those metro networks, we have simulated 28 Barbási-Albert networks[10]. Those BA networks are simulated with the same sizes as those 28 metro networks. Figure 2 gives the largest connected components G of metro networks and

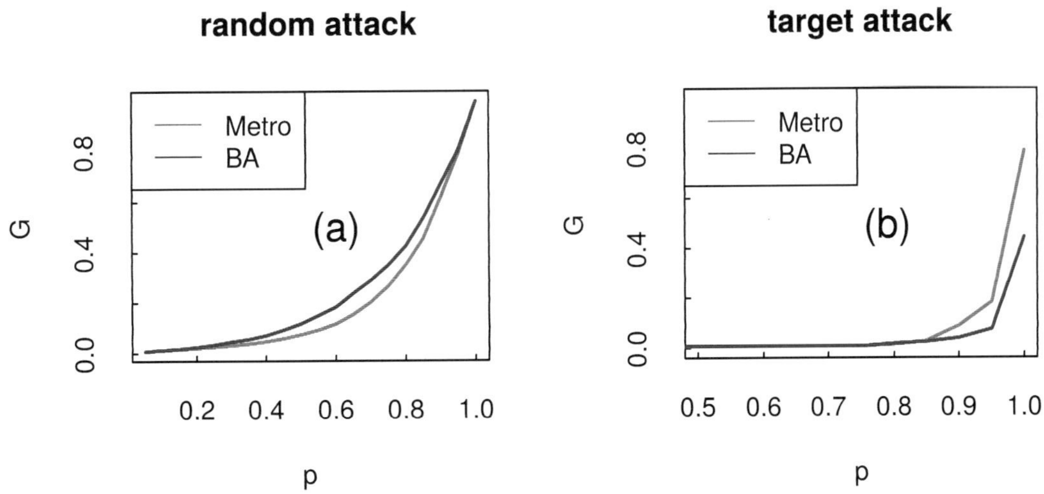

Figure 2. The comparison between metro networks and BA networks under two attack strategies. The red line is the average largest connected component for metro networks and the blue line is the average largest connected component for BA networks.

the largest connected components G of the corresponding BA networks under both random attack and target attack. In figure 2(a) we can see that average size of giant components for BA networks are larger than that of metro networks under random attack. Whilst in figure 2(b) the size of giant component for BA networks are smaller than that of metro networks. The different behavior of metro networks compare to BA networks under different attack strategies means that the design of metro networks is more robust under target attack.

Meanwhile, we also simulated 28 ER networks with the same size as those 28 metro networks[7]. Figure 3 gives the giant component of both metro networks and ER networks under two attack strategies. We observe that the size of giant component for metro networks are consistently smaller than that of ER networks for different remaining node fraction p. Thus we conclude that the metro networks are less robust compare with ER networks under both random and target attacks. This maybe rooted in the spatial nature of metro networks which will hinder the formation of shortcuts during the growth of metro networks. Thus the lack of shortcuts and long range connections make the metro networks extremely vulnerable compare to random networks.

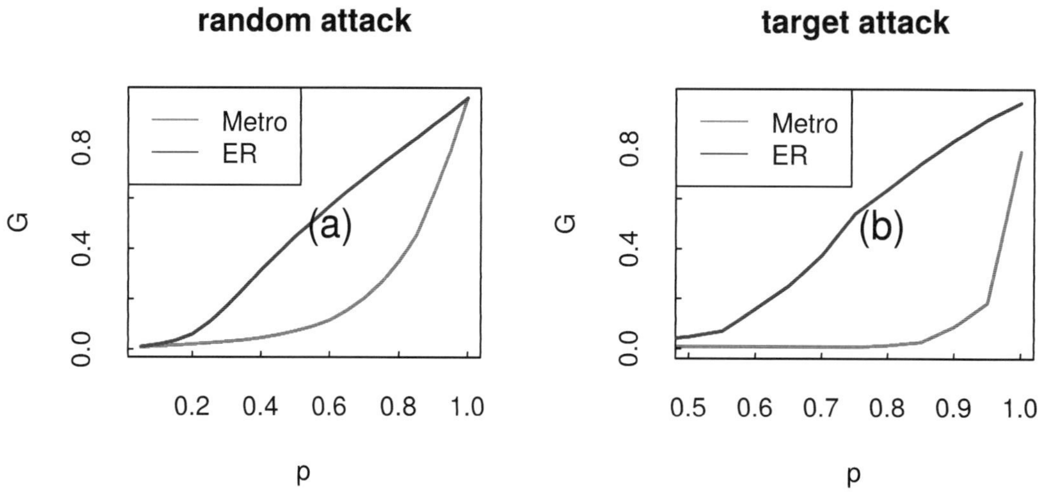

Figure 3. The comparison between metro networks and ER networks under two attack strategies. The red line is the average largest connected component for metro networks and the blue line is the average largest connected component for ER networks.

4. Conclusion

In conclusion, we have analyzed the robustness of 28 worldwide metro networks. By using two attack strategies: random attack and target attack, we can access the robustness of metro networks from the size of largest connected component. We find that the metro networks are more robust under random attack, but fragile under target attack. This gives a hint about the importance of those hub stations which deserve strict protection.

We also simulate 28 BA networks and 28 ER networks with the same sizes of corresponding metro networks. Under two different attack strategies, we find metro networks are more robust than BA networks under target attack and less robust than BA networks under random attack. Meanwhile, the metro networks have been proved to be less robust than ER networks under both random and target failures.

Those results presented here can be recognized as a guidance for future metro system design. There also exist some research directions such as robustness evaluation of metro networks from network control, the robustness of metro networks coupled with other infrastructure systems, e.g., bus networks and railway networks. It is also very interesting to access the robustness of metro networks from their community structure[11, 12, 13]. Those will subject to further researches.

Acknowledgement

This work is supported in part by the Programme of Introducing Talents of Discipline to Universities under grant NO. B08033 and the program of China Scholarship Council (No. 201606770023).

References

[1] Barabási A L 2011 *Nature Physics* **8** 14–16
[2] Newman M E J 2003 *SIAM Rev* **45** 167

[3] Boccaletti S, Latora V, Moreno Y, Chavez M and Hwang D U 2006 *Physics Reports* **424** 175–308
[4] Zhao L, Li W and Cai X 2016 *Physics Letters A* **380** 654–666
[5] Zhao L, Wang G J, Wang M, Bao W, Li W and Stanley H E 2018 *Physica A, https://doi.org/10.1016/j.physa.2018.05.039.*
[6] Zhao L, Li W, Fenu A, Podobnik B, Wang Y and Stanley H E 2018 *Journal of Statistical Mechanics: Theory and Experiment* **2018** 23402
[7] Erdös P and Rényi A 1959 *Publicationes Mathematicae (Debrecen)* **6** 290–297
[8] Barabási A L and Albert R 1999 *Science* **286** 509–512
[9] Li W, Gu J, Liu S, Zhu Y, Deng S, Zhao L, Han J and Cai X 2014 *Preprint* 1403.7844
[10] Albert R and Barabasi A L 2002 *Reviews of Modern Physics* **74** 47–97
[11] Fortunato S and Hric D 2016 *Physics Reports* **659** 1–44
[12] Han J, Li W, Su Z, Zhao L and Deng W 2016 *The European Physical Journal B* **89** 272
[13] Han J, Li W, Zhao L, Su Z, Zou Y and Deng W 2017 *PLoS ONE* **12** e0188655

Critical behavior of a generalized Bak-Sneppen model

Jihui Han[1], **Wei Li**[2] **and Weibing Deng**[2]

[1] School of Computer and Communication Engineering, Zhengzhou University of Light Industry, Zhengzhou, China

[2] Key Laboratory of Quark and Lepton Physics (MOE) and Institute of Particle Physics, Central China Normal University, Wuhan, China

E-mail: hanjihui@zzuli.edu.cn (Jihui Han), liw@mail.ccnu.edu.cn (Wei Li), wdeng@mail.ccnu.edu.cn (Weibing Deng)

Abstract. We propose an alternative to the Bak-Sneppen model for species co-evolution. In our model the closest neighbors of the least adapted species are replaced by new species with a certain probability α. The probability α can be regarded as the interaction strength between nearest species. We show that the system can always self-organizes to a critical state when the interaction strength α is larger than zero. The critical exponents of our model are estimated through the finite size scaling analysis, and found to be the same as those of the original Bak-Sneppen model. We also find that the critical dimension of our model is 4, below which avalanches are compact and above which the critical behavior is mean-field like. Moreover, a strong interaction strength leads to a lower critical fitness.

1. Introduction

Self-organized criticality (SOC) [1] has become one of the most studied concepts of non-equilibrium statistical mechanics. It can be used to explain the origin of the ubiquitous occurrence of fractal structures in nature [2], noise with $1/f$ power spectrum [1] and punctuated equilibrium [3].

Among many SOC models, the Bak-Sneppen (BS) model [4], which mimics co-evolution between interacting species, is the simplest one. In this model, an ecosystem is characterized by L^d species on a d-dimensional lattice of linear size L. A random number called fitness value f_i, drawn from a uniform distribution between 0 and 1, is assigned to each site of the lattice as the initial state of the system. At each time step, the site with the smallest fitness and its $2d$ nearest neighbors are replaced by new random numbers drawn from the same distribution. After some transient time, which depends on the system size, the model can reach a statistically stationary state where the density of fitness is uniformly distributed between f_c and 1, and vanishes under this critical fitness. In the stationary state, the model exhibits punctuated equilibrium and the complexity of this regime can be revealed by the existence of spatio-temporal power-law distribution of avalanches.

The BS model has received much attention [3, 5] in the statistical physics community and has been studied through various approaches, including numerical simulation [6, 7], theoretical analysis [8], and mean-field theory [9]. Many variants of the Bak-Sneppen model have been proposed, such as (1) the discrete Bak-Sneppen model, in which fitnesses are only take the values 0 and 1; (2) the anisotropic Bak-Sneppen (aBS) model [10], in which only the least fit species and its right-hand nearest neighbor are mutated; (3) the random neighbors BS model

[11], in which the least fit species and $K-1$ randomly chosen other ones are replaced by new species.

In the BS model, the least fit species and its $2d$-nearest neighbors have full interaction, and are replaced by new species at the same time. This is not true in most real situations. In Ref. [12], the authors distinguish the strength of interactions of the BS model in one dimensional lattice. They found that as long as the interaction strength is non-zero, the model can always evolve to a critical state where the avalanche sizes subject to a power-law distribution. The critical exponent τ for the avalanche life distribution slightly relies on α. The critical exponent γ for average avalanche size distribution and the avalanche fractal dimension D increase slowly as α increases. Here, we focus on the BS model with variable interaction strength in high dimensions. We implemented extensive simulations, estimated critical exponents and critical fitness accurately, and verified the scaling relations which are derived from the BS model are still valid in our model.

2. The model description

Our model is defined as follows:

(1) Initially, on a d-dimensional lattice with linear length L, a random number called fitness drawn from distribution $P(f)$, is assigned to each site.

(2) At each time step, the global minimum fitness is located, and replaced by a new random number drawn from the same distribution $P(f)$. Its $2d$-nearest neighbors are replaced by new random numbers also taken from $P(f)$ with probability α. Here, $P(f)$ is a uniform distribution between 0 and 1.

(3) Repeat Step (2)

As mentioned above, α is interpreted as interaction strength, it is fixed during the system evolution and is independent of species' fitness. The value of α ranges from 0 to 1. If α is set to 0, there is no interaction and all fitness values will eventually become 1 where no SOC can be observed. If α is set to 1, the original BS model is restored.

3. Simulation results

We use a generalized branching process [3] to describe an f_0 avalanche, which provides an efficient way to study the f_0 avalanches and is completely free from system size corrections . In simulations of this branching process, to describe an individual avalanche, we can assume that it started at the origin, and all times we need only keep track of those numbers below f_0 with explicit lattice reference. We have first used this simulated branching process to extrapolate threshold values f_c for $1 \leq d \leq 8$ with $\alpha = 0.1, 0.2, \cdots, 1$. Then run up to 4×10^6 critical avalanches to determine their statistical properties.

3.1. Critical exponent of average avalanche size γ distribution and critical fitness f_c

Normally, we need to observe the avalanches based on f_c. However, the critical fitness cannot be precisely identified. An alternative way is to choose an auxiliary parameter f_0 very close to f_c and observe the so-called f_0-avalanche near the critical point. From the hierarchical structure of the f_0-avalanche, the following exact differential equation for the average size of f_0 avalanches can be derived [3]

$$\frac{d \ln \langle S \rangle_{f_0}}{d f_0} = \frac{\langle n_{cov} \rangle_{f_0}}{1 - f_0}, \tag{1}$$

where S is the avalanche size, $\langle n_{cov} \rangle_{f_0}$ is the average number of sites covered by f_0-avalanches, f_0 is the auxiliary parameter to define an f_0-avalanche. With the assumption that when

f_0 approaches the critical value f_c, the average size of avalanches diverges as $(f_c - f_0)^{-\gamma}$. Substituting this scaling ansatz into Eq. (1) gives the so-called γ equation [3]

$$\gamma = \lim_{f_0 \to f_c} \frac{\langle n_{cov} \rangle_{f_0} (f_c - f_0)}{1 - f_0}, \tag{2}$$

where γ is the critical exponent of average avalanche size distribution, and f_c is the critical fitness. Note that γ appears as a constant for a certain model [3]. So we can plot $(1 - f_0)/\langle n_{cov} \rangle_{f_0}$ for different values of f_0, as shown in Figure 1, the intersection with the vertical axis gives f_c very accurately, and the asymptotic slope gives γ.

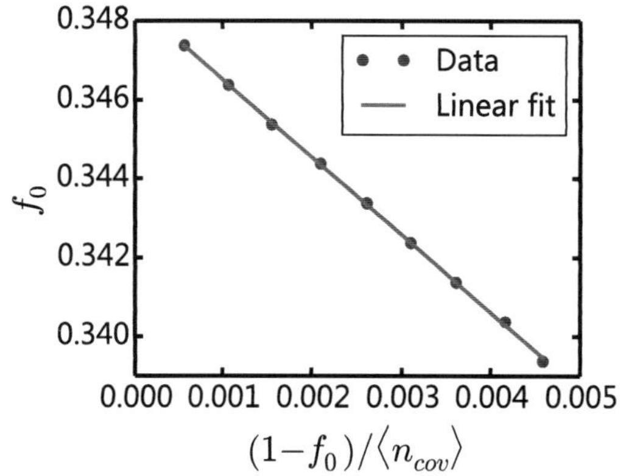

Figure 1. f_0 as a function of $(1 - f_0)/\langle n_{cov} \rangle$ of our model in two-dimensional case with $\alpha = 0.9$. The solid line is a least square fit of the data. The intersection with the vertical axis gives $f_c = 0.348474(3)$, and the slope gives $\gamma = 1.69(1)$.

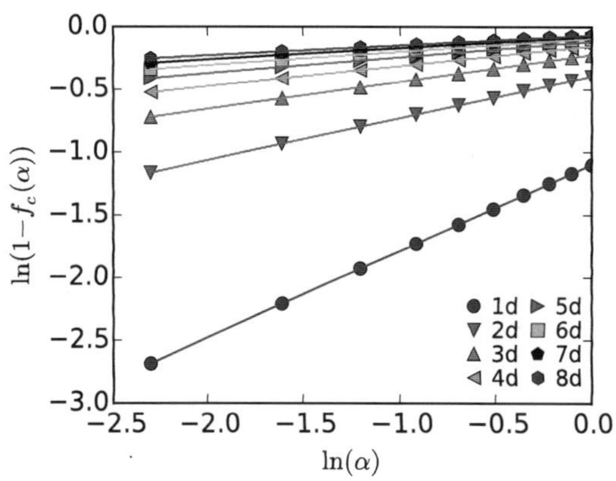

Figure 2. $\ln(1 - f_c(\alpha))$ as a function of $\ln(\alpha)$ for our model in dimensions 1 to 8. The solid lines are least square fits of the data. The intersection with the vertical axis gives $\ln(1 - fc(1))$. Clearly, the critical fitness depends heavily on the underlying interaction details and have a simple relation with α: $f_c(\alpha) = 1 - C\alpha^{\beta}$, where $C = 1 - f_c(1)$.

We have estimated γ and f_c for our model with $\alpha = 0.1, 0.2, \cdots, 1.0$ in dimensions 1 to 8. We found that f_c depends heavily on α, but the critical exponent γ only depend on dimensionality and are entirely insensitive to the changes in α. Fig. 2 shows $\ln(1-f_c(\alpha))$ as a function of $\ln(\alpha)$ for our model in different dimensions. We found that the critical fitness and the interaction strength has a simple relation: $f_c(\alpha) = 1 - C\alpha^{\beta}$, with different values of β in different dimensions, where $C = 1 - f_c(1)$. In a certain dimension, a stronger interaction strength leads to a lower critical fitness. The values of critical exponent γ for our model in dimensions 1 to 8 are listed in Table 1. As can be seen, γ does not change when $d > 4$, which is in line with the result in Ref. [13].

3.2. Scaling function and data collapse

The statistics of the avalanches clearly depends on the value of f_0. The larger it is, the larger the expected size of the avalanche is. We use the following scaling ansatz for the probability distribution $P(s, f_0)$ and complementary cumulative probability distribution $P(S \geq s, f_0)$ of f_0 avalanches of size s,

$$P(s, f_0) = s^{-\tau} g\left(s (f_c - f_0)^{1/\sigma}\right), \tag{3a}$$

$$P(S \geq s, f_0) = s^{-\tau+1} h\left(s (f_c - f_0)^{1/\sigma}\right). \tag{3b}$$

Here τ and σ are model dependent exponents. $g(x)$ and $h(x)$ are scaling functions, which decays rapidly to zero when $x \gg 1$ and goes to a constant when $x \to 0$. This ansatz has been confirmed by numerical simulations in Refs. [4, 14–21]. When f_0 is lowered below f_c, the avalanche distribution acquire a cutoff $S_{co} = (f_c - f_0)^{-1/\sigma}$.

From Eq (3b) we can see that two completely independent exponents combined in a nontrivial way to form a single one leads to an enormous simplification in the description of the model. The values of $P(S \geq s, f_0)$ for various s and f_0 can be made to collapse on a single curve if $s^{\tau-1}P(S \geq s, f_0)$ is plotted against $s(f_c - f_0)^{1/\sigma}$. In order to remove the subjectiveness of the data collapse we adopt a method from [22], in which a measure is proposed to quantify the 'goodness of collapse'. The measure P_b in our scenario is defined as,

$$P_b = \left[\frac{1}{N_{over}} \sum_{p} \sum_{j \neq p} \sum_{i,over} \left| s_{i,j}^{\tau-1} P_{i,j} - \varepsilon_p \left(s_{i,j} (f_c - f_{0j})^{1/\sigma}\right)\right|^q \right]^{1/q}, \tag{4}$$

where $\varepsilon_p(x)$ is the interpolating function based on the values of set p bracketing the argument in question of set j. Different f_0 leads to different data sets. $s_{i,j}$ is the ith avalanche size for the jth set of f_0. $P_{i,j}$ is the ith complementary cumulative avalanche distribution probability for jth set (i.e. $P_{i,j} = P(S \geq s_i, f_{0j})$). Here we use the complementary cumulative avalanche distribution $P(S \geq s, f_0)$ instead of the avalanche distribution $P(s, f_0)$ because the statistics of the former is much better than latter (binned data will lose a lot of information). The innermost sum over i is done only over the overlapping regions (denoted by the 'i,over'), i.e. only those values of set j are considered for which $x_{i,j} = s_{i,j}(f_c - f_{0j})^{1/\sigma}$ belongs to the interval spanned by the corresponding x-values of set p. N_{over} is the total number of such pairs. Here, we use $q = 1$ and a cubic B-spline interpolation for $\varepsilon_p(x)$. A minimization of P_b over (τ, σ) can be used to extract the optimal values of the exponents. Estimates of errors can be obtained from the width of the minimum. From an expansion of $\ln P_b$ around the minimum at (τ_0, σ_0), the width is estimated as:

$$\Delta\tau = \eta\tau_0 \left[2 \ln \frac{P_b(\tau_0 \pm \eta\tau_0, \sigma_0)}{P_b(\tau_0, \sigma_0)}\right]^{-1/2} \tag{5a}$$

$$\Delta\sigma = \eta\sigma_0 \left[2 \ln \frac{P_b(\tau_0, \sigma_0 \pm \eta\sigma_0)}{P_b(\tau_0, \sigma_0)}\right]^{-1/2} \tag{5b}$$

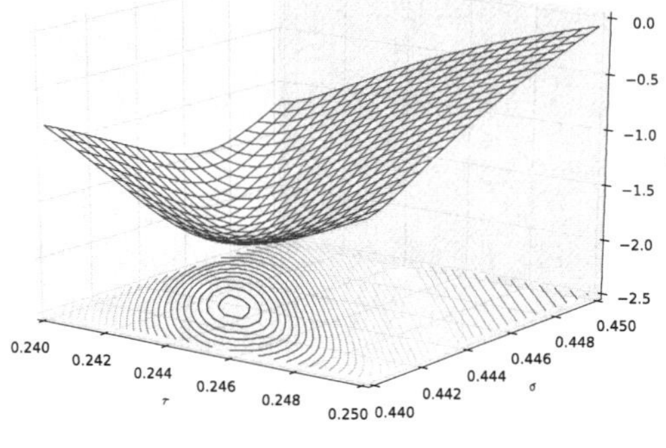

Figure 3. The measure P_b with $q = 1$ for our model in two-dimensional case is shown over the (τ, σ) plane. The z-axis is in log-scale. A few contours of constant $\ln(P_b)$ are shown by projecting the surface on the (τ, σ) plane. Here, $\alpha = 0.5$, $f_c - f_0 = 0.001, 0.002, \cdots, 0.01$.

As can be seen, this method not only can be used to measure the goodness of data collapse but also can estimate the exponents τ and σ all at once. We also estimate τ by fitting the avalanche data using methods in Ref. [23], and the results are consistent with those of the data collapse method.

We counted 2×10^6 avalanches for various values of f_0 in each dimension with $\alpha = 0.1, 0.2, \cdots, 1.0$. Fig. 3 shows the measure of data collapse P_b of our model in two-dimensional case with $\alpha = 0.5$. A minimization of P_b gives us the exponents $\tau = 1.244$, $\sigma = 0.443$, with $P_b = 0.18$. We also estimate τ and σ for different values of α in dimensions 1 to 8, and find that the exponents depend only on the dimensionality (see Table 1). Fig. 4 shows the data collapse of our model in two-dimensional case with $\alpha = 0.5$ and 1. So increasing the interaction strength α modifies the critical fitness f_c but not the critical exponents, which is consistent with results in Ref. [24]. The critical fitness f_c depends heavily on the underlying lattice details, while the critical exponents only depend on the dimensionality and are entirely insensitive to the underlying lattice details.

Table 1. List of the critical exponents of our model, for $1 \le d \le 8$. The errors in the exponent are less than 1%.

d	1	2	3	4	5	6	7	8
τ	1.08	1.25	1.37	1.46	1.50	1.50	1.50	1.50
σ	0.34	0.44	0.47	0.46	0.50	0.50	0.50	0.50
γ	2.69	1.69	1.33	1.17	1.00	1.00	1.00	1.00
D	2.43	2.92	3.35	3.92	4.00	4.00	4.00	4.00
μ	0.41	0.68	0.91	0.99	1.00	1.00	1.00	1.00
d_f	1	2	3	4	4	4	4	4

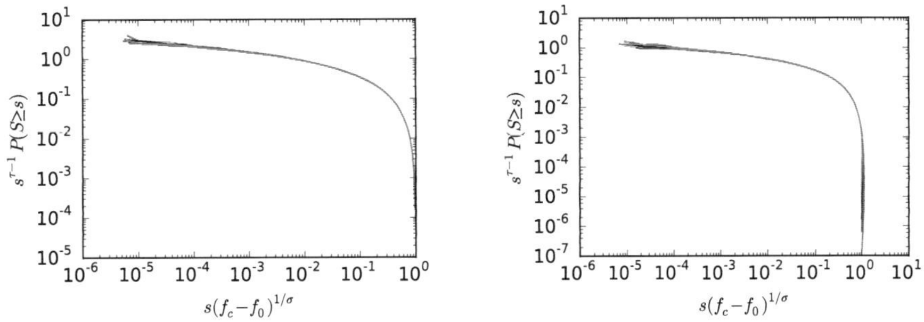

Figure 4. Data collapse of our model for $f_c - f_0 = 0.001, 0.002, \cdots, 0.01$, with $\alpha = 0.5$ (left) and 1 (right) in two-dimensional case, using the best estimate exponents $\tau = 1.244$, $\sigma = 0.443$.

3.3. Fractal of avalanches

We recorded the instances of having activity at a distance r relative to the origin at update step s. Similar to a random walk, the moments of the distribution define the avalanche dimension exponent D via $\langle r\rangle_s \sim s^{1/D}$. We estimated D by least square fit $\ln(s)$ and $\ln(\langle r\rangle_s)$ in each dimension with different α, which shows that D is insensitive to changes in α. Fig. 5 shows s vs $\langle r\rangle$ of our model in two-dimensional case with $\alpha = 0.5$. We used the value of our best fit for the exponent D as given for each dimension in Table 1. For all dimensions $s/\langle r\rangle^D$ tends to be stable for increasing $\langle r\rangle$. For $d \geq 4$, the fractal exponent D corresponds to the mean-field result [25, 26]

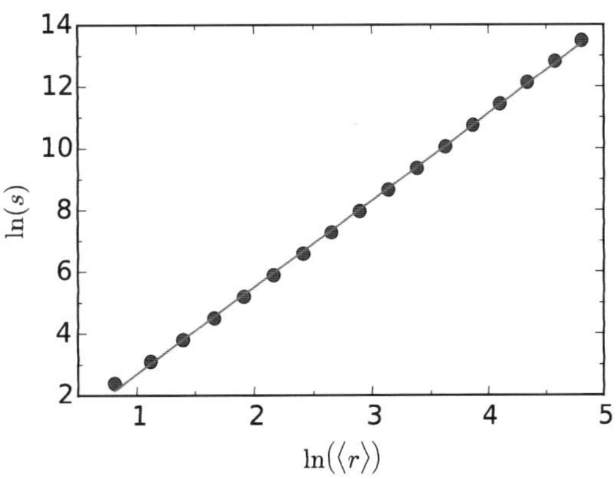

Figure 5. $\ln(s)$ as a function of $\ln(\langle r\rangle)$ of our model in two-dimensional case with $\alpha = 0.5$. The solid line is a least square fit of the data and the slope gives the fractal exponent $D = 2.92$.

We also measured the scaling of the coverage with the lifetime of an avalanche $\langle n_{cov}\rangle \sim s^{\mu}$ and listed our best fit for the exponent μ in Table 1. Fig. 6 shows $\langle n_{cov}\rangle$ as a function of s of our model in two-dimensional case with $\alpha = 0.5$.

To test whether the avalanche is compact or not in a certain dimension, we estimated exponent d_f defined by $\langle n_{cov}\rangle \sim R^{d_f}$, where R is the radius of gyration of a avalanche covered domain. Avalanches are fractal if $d_f < d$ and compact if $d_f = d$. Fig. 7 shows the plot of $\langle n_{cov}\rangle$ as a function of R of our model in two-dimensional case with $\alpha = 0.5$. Since there should be only one characteristic length for a compact avalanche, there should have $\langle r\rangle \sim R$, and thus, $\mu = d/D$

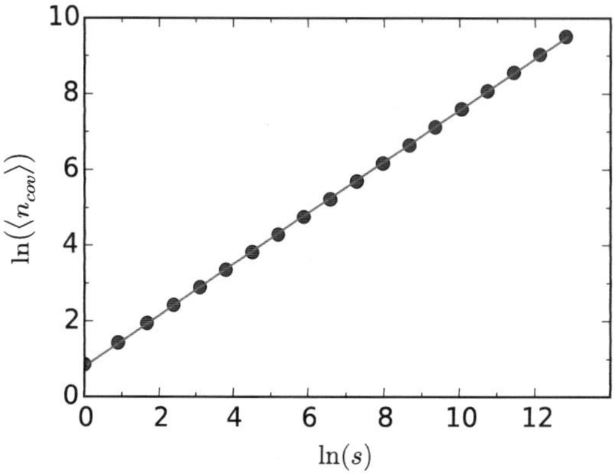

Figure 6. $\ln(\langle n_{cov} \rangle)$ as a function of $\ln(s)$ of our model in two-dimensional case with $\alpha = 0.5$. The solid line is a least square fit of the data and the slope gives exponents $\mu = 0.68$.

for a compact avalanche, and $\mu = 1$ in the mean-field limit. Clearly, the exponents in Table 1 for $d \leq 4$ in line with $\mu = d/D$, and for $d = 5, 6, 7, 8$ are in perfect agreement with mean-field values. That is to say, avalanches of our model are compact until $d = 4$, which is consistent with [13]. Because of the coincidence of the exponents of our model with those of the BS model, the scaling relations in [3] are preserved. We can easily verify the scaling relations of exponents in Table 1.

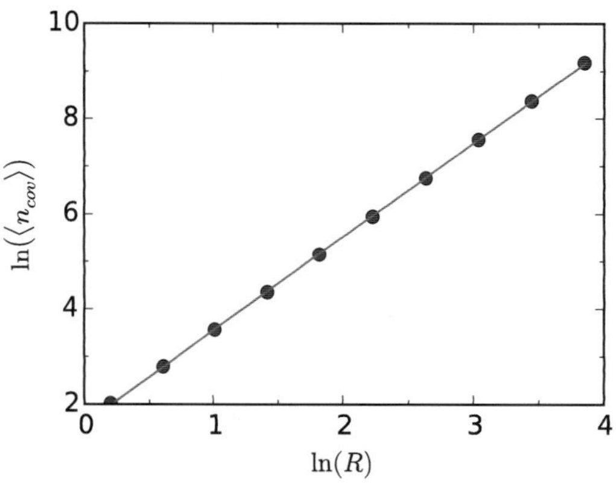

Figure 7. $\ln(\langle n_{cov} \rangle)$ as a function of $\ln(R)$ of our model in two-dimensional case with $\alpha = 0.5$. The solid line is a least square fit of the data and the slope gives exponents $d_f = 2.0$.

4. Conclusions

To summarize, we have studied a stochastic Bak-Sneppen model in dimensions 1 to 8 by extensive numerical simulations. The avalanches are regarded as a branching process, and the branching factor $r_b = 1 + 2d\alpha$, as long as $\alpha > 0$, the branching factor $r_b > 1$ in all case, and the nontrivial SOC states are observed in every dimension. The critical fitness changes with α according to $f_c(\alpha) = 1 - C\alpha^{\beta}$, with different β in different dimensions, where $C = 1 - f_c(1)$. The critical exponents of our model do not change with α, and are consistent with those of the origin BS model.

Acknowledgments
This work was supported in part by National Natural Science Foundation of China (Grant No. 11505071, Recipient: Weibing Deng) and Program of Introducing Talents of Discipline to Universities (Grant No. B08033, Recipient: Wei Li).

References
[1] Bak P, Tang C and Wiesenfeld K 1987 *Phys. Rev. Lett.* **59**(4) 381–384 URL http://link.aps.org/doi/10.1103/PhysRevLett.59.381
[2] Bardeen J, Cooper L N and Schrieffer J R 1957 *Phys. Rev.* **108** 1175
[3] Paczuski M, Maslov S and Bak P 1996 *Phys. Rev. E* **53**(1) 414–443 URL http://link.aps.org/doi/10.1103/PhysRevE.53.414
[4] Bak P and Sneppen K 1993 *Physical Review Letters* **71** 4083–4086 ISSN 00319007
[5] Klimek P, Thurner S and Hanel R 2010 *Physical Review E - Statistical, Nonlinear, and Soft Matter Physics* **82** ISSN 15393755
[6] Grassberger P 1995 *Physics Letters A* **200** 277 – 282 ISSN 0375-9601 URL http://www.sciencedirect.com/science/article/pii/0375960195001797
[7] Paczuski M, Maslov S and Bak P 1996 *Phys. Rev. E* **53**(1) 414–443 URL https://link.aps.org/doi/10.1103/PhysRevE.53.414
[8] Maslov S 1996 *Physical Review Letters* **77** 1182–1185 ISSN 0031-9007 URL http://link.aps.org/doi/10.1103/PhysRevLett.77.1182
[9] Marsili M, De Los Rios P and Maslov S 1998 *Physical Review Letters* **80** 1457–1460 ISSN 0031-9007 URL http://link.aps.org/doi/10.1103/PhysRevLett.80.1457
[10] Maslov S, De Los Rios P, Marsili M and Zhang Y C 1998 *Phys. Rev. E* **58**(6) 7141–7145 URL http://link.aps.org/doi/10.1103/PhysRevE.58.7141
[11] Cafiero R, De Los Rios P, Valleriani A and Vega J L 1999 *Phys. Rev. E* **60**(2) R1111–R1114 URL http://link.aps.org/doi/10.1103/PhysRevE.60.R1111
[12] Li W, Luo Y, Wang Y and Cai A 2011 *Chinese Science Bulletin* **56** 3639–3642 ISSN 1001-6538 URL http://dx.doi.org/10.1007/s11434-011-4654-1
[13] Boettcher S and Paczuski M 2000 *Phys. Rev. Lett.* **84**(10) 2267–2270 URL http://link.aps.org/doi/10.1103/PhysRevLett.84.2267
[14] Ray T S and Jan N 1994 *Phys. Rev. Lett.* **72**(25) 4045–4048 URL http://link.aps.org/doi/10.1103/PhysRevLett.72.4045
[15] de Boer J, Derrida B, Flyvbjerg H, Jackson A D and Wettig T 1994 *Phys. Rev. Lett.* **73**(6) 906–909 URL http://link.aps.org/doi/10.1103/PhysRevLett.73.906
[16] de Boer J, Jackson A D and Wettig T 1995 *Phys. Rev. E* **51**(2) 1059–1074 URL http://link.aps.org/doi/10.1103/PhysRevE.51.1059
[17] Tang L H and Leschhorn H 1993 *Phys. Rev. Lett.* **70**(24) 3832–3832 URL http://link.aps.org/doi/10.1103/PhysRevLett.70.3832
[18] Leschhorn H and Tang L H 1994 *Phys. Rev. E* **49**(2) 1238–1245 URL http://link.aps.org/doi/10.1103/PhysRevE.49.1238
[19] Martys N, Robbins M O and Cieplak M 1991 *Phys. Rev. B* **44**(22) 12294–12306 URL http://link.aps.org/doi/10.1103/PhysRevB.44.12294

[20] Bruinsma R and Aeppli G 1984 *Phys. Rev. Lett.* **52**(17) 1547–1550 URL http://link.aps.org/doi/10.1103/PhysRevLett.52.1547

[21] Ji H and Robbins M O 1992 *Phys. Rev. B* **46**(22) 14519–14527 URL http://link.aps.org/doi/10.1103/PhysRevB.46.14519

[22] Bhattacharjee S M and Seno F 2001 *Journal of Physics A: Mathematical and General* **34** 6375 URL http://stacks.iop.org/0305-4470/34/i=33/a=302

[23] Clauset A, Shalizi C R and Newman M E J 2009 *SIAM Review* **51** 661–703 (*Preprint* https://doi.org/10.1137/070710111) URL https://doi.org/10.1137/070710111

[24] Garcia G J and Dickman R 2004 *Physica A: Statistical Mechanics and its Applications* **342** 516 – 528 ISSN 0378-4371 URL http://www.sciencedirect.com/science/article/pii/S0378437104008738

[25] Boettcher S and Paczuski M 1996 *Phys. Rev. Lett.* **76**(3) 348–351 URL http://link.aps.org/doi/10.1103/PhysRevLett.76.348

[26] Boettcher S and Paczuski M 1996 *Phys. Rev. E* **54**(2) 1082–1095 URL http://link.aps.org/doi/10.1103/PhysRevE.54.1082

IOP Conf. Series: Journal of Physics: Conf. Series **1113** (2018) 012012 doi:10.1088/1742-6596/1113/1/012012

On quantum-mechanical origin of statistical mechanics

Yuho Yokoi[1] **and Sumiyoshi Abe**[2, 3, 4]

[1] Graduate School of Engineering, Mie University, Mie 514-8507, Japan
[2] Physics Division, College of Information Science and Engineering, Huaqiao University, Xiamen 361021, China
[3] Department of Physical Engineering, Mie University, Mie 514-8507, Japan
[4] Institute of Physics, Kazan Federal University, Kazan 420008, Russia

E-mail: yuho_451@yahoo.co.jp (Y. Yokoi), suabe@sf6.so-net.ne.jp (S. Abe)

Abstract. The problem of deriving statistical mechanics from the stationary Schrödinger equation is discussed. The interaction Hamiltonian, which dynamically induces entanglement of the specific type, is constructed in a unified way based on the gauge principle. It is shown how microcanonical ensembles in both Bose-Einstein and Fermi-Dirac statistics emerge in the vanishing-interaction limit.

In spite of its great success for about 1.5 centuries, classical statistical mechanics still remains challenging at the level of its fundamental principles. It is of significance to view them from the standpoint of quantum mechanics. There are a series of indications, which suggest that quantum mechanics may serve classical statistical mechanics with its foundations. The ergodic hypothesis is yet to be generically proved. The probabilistic concept in classical statistical mechanics is due to lack of knowledge about microscopic motion of a large number of particles, whereas in quantum mechanics it is a law of nature. Also, the Gibbs factor could be understood in such a way that identical particles are indistinguishable even classically, but it automatically follows from quantum mechanics. Moreover, the Planck constant is indispensable in order for the entropy to be definable in phase space. The thermal wavelength cannot be given its expression without the Planck constant. The quantum-classical correspondence in statistical mechanics is concerned not with the Planck constant but with the temperature.

Recently, a number of efforts have been made on understanding of statistical mechanics based on quantum mechanics. There, the concepts of ensemble typicality [1, 2] and eigenstate thermalization [3, 4] have been playing vital roles. In addition to these, we have recently developed a discussion [5] (see also Refs. [6, 7, 8]) about derivations of Bose-Einstein and Fermi-Dirac statistics from the stationary Schrödinger equation. In this article, we would like to summarize the discussion in a further unified way. We show how microcanonical ensembles in those statistics emerge from quantum mechanics.

In order to derive microcanonical ensemble from quantum mechanics, the following two issues are required in Ref. [5]: (I) perfect decoherence in terms of the energy eigenbasis should be realized for an *isolated* system, and (II) the principle of equal *a priori* probability has to be consistent with the solution of the stationary Schrödinger equation. The latter forces an interaction between particles to entail a specific pattern of entanglement in the quantum state as the solution. It should be noted that weak inter-particles interactions are in fact needed for deriving statistical mechanics even of the ideal gas: without interactions, a system could never relax to the equilibrium state. However, such interactions can be ignored when the description of the system shifts from the mechanical to the statistical mechanical.

The system we consider here consists of N identical particles with the common energy quantum ε. Its free Hamiltonian has the form

$$\hat{H}_0 = \varepsilon \sum_{i=1}^{N} \hat{n}_i. \tag{1}$$

$\hat{n}_i \equiv \hat{a}_i^\dagger \hat{a}_i$ is the number operator of the ith particle, where $\hat{a}_i^\dagger$ and $\hat{a}_i$ are the creation and annihilation operators satisfying the algebra: $\left[\hat{a}_i, \hat{a}_j^\dagger\right]_\pm = \delta_{ij}$, $[\hat{a}_i, \hat{a}_j]_\pm = \left[\hat{a}_i^\dagger, \hat{a}_j^\dagger\right]_\pm = 0$ with the notation $\left[\hat{A}, \hat{B}\right]_\pm = \hat{A}\hat{B} \pm \hat{B}\hat{A}$. The commutators (anticommutators) correspond to the bosonic (fermionic) particles. As mentioned above, it is necessary to introduce interactions between the particles. Since perfect decoherence should be realized for an isolated system as in the requirement (I), the interactions may be relevant to phases that are (approximately) conjugate to the number of the particles whose eigenstates form the energy eigenbasis. In Ref. [5], the unitary phase operator widely used in quantum optics is employed. It is defined as follows [9, 10]:

$$\exp(\mathrm{i}\hat{\phi}_i) = \sum_{m_i=0}^{s} \exp(\mathrm{i}\theta_{m_i}) \left|\theta_{m_i}\right\rangle_i {}_i\!\left\langle\theta_{m_i}\right|. \tag{2}$$

Here, $\left|\theta_{m_i}\right\rangle_i$ is the discrete phase state of the ith particle given by $\left|\theta_{m_i}\right\rangle_i = (s+1)^{-1/2} \sum_{n_i=0}^{s} \exp(\mathrm{i} n_i \theta_{m_i}) \left|n_i\right\rangle_i$ with the normalized number state $\left|n_i\right\rangle_i = \left(\hat{a}_i^\dagger\right)^{n_i} \left|0\right\rangle_i / \sqrt{n_i!}$ satisfying $\hat{n}_i \left|n_i\right\rangle_i = n_i \left|n_i\right\rangle_i$ $(n_i = 0, \ldots, s)$, where $\left|0\right\rangle_i$ is the normalized ground state annihilated by $\hat{a}_i$: $\hat{a}_i \left|0\right\rangle_i = 0$. In the bosonic case, s is a large positive integer, implying that the phase state is defined in the $(s+1)$-dimensional truncated space, and the limit $s \to \infty$ has to be taken after all calculations about the phase operators and phase states [9, 10]. On the other hand, s is fixed to be $s = 1$ for the fermions. θ_{m_i} is the discrete phase eigenvalue given by $\theta_{m_i} = 2\pi m_i/(s+1)$ $(m_i = 0, \ldots, s)$. Both $\{\left|n_i\right\rangle_i\}_{n_i}$ and $\left\{\left|\theta_{m_i}\right\rangle_i\right\}_{m_i}$ form the complete orthonormal systems in the $(s+1)$-dimensional space. Thus, $\left|\theta_{m_i}\right\rangle_i$ is the eigenstate of the phase operator in Eq. (2): $\exp(\mathrm{i}\hat{\phi}_i) \left|\theta_{m_i}\right\rangle_i = \exp(\mathrm{i}\theta_{m_i}) \left|\theta_{m_i}\right\rangle_i$. As known, the phase operator cannot strictly be conjugate to the number operator since the spectrum of the number operator is discrete. The related discussions date back to Dirac [11], and one can see the phase-operator problem as a prototype of anomaly [12].

In terms of the basis $\{\left|n_i\right\rangle_i\}_{n_i}$, the phase operator in Eq. (2) is expressed as follows:

$$\exp(\mathrm{i}\hat{\phi}_i) = \sum_{n_i=0}^{s-1} \left|n_i\right\rangle_i {}_i\!\left\langle n_i + 1\right| + \left|s\right\rangle_i {}_i\!\left\langle 0\right|. \tag{3}$$

What is of crucial importance for the subsequent discussion is the existence of the hidden gauge-theoretical structure. From Eq. (3), it is clear that under the c-number local (i.e., n_i-dependent) gauge transformation

$$\left|n_i\right\rangle_i \to \left|n_i\right\rangle_i \exp(\mathrm{i}\Lambda_{n_i,\mu_i}), \quad \Lambda_{n_i,\mu_i} = n_i \theta_{\mu_i} = \frac{2\pi\mu_i n_i}{s+1} \quad (\mu_i = 0, \ldots, s), \tag{4}$$

$\hat{\phi}_i$ transforms as the gauge field

$$\hat{\phi}_i \to \hat{\phi}_i - \partial\Lambda_{n_i,\mu_i}, \tag{5}$$

where

$$\partial\Lambda_{n_i,\mu_i} \equiv \Lambda_{n_i+1,\mu_i} - \Lambda_{n_i,\mu_i} = \theta_{\mu_i}. \tag{6}$$

This structure turns out to play a central role for constructing the interaction Hamiltonian.

Let us define the following operator:

$$\hat{v}_i = \left[\exp(\mathrm{i}\hat{\phi}_i) - |s\rangle_{i\,i}\langle 0|\right] \exp(-\mathrm{i}\theta_{m_i}) q^{\hat{F}_i}, \tag{7}$$

where

$$\hat{F}_i = \sum_{j<i} \hat{n}_j \qquad \left(\hat{F}_1 \equiv 0\right), \tag{8}$$

and $q = 1\ (-1)$ for the bosons (fermions). The factor $q^{\hat{F}_i}$ is identically equal to unity for the bosons, whereas it gives rise to nontrivial sign changes for the fermions. Consider a generic number state of the fermions: $|n_1, n_2, \ldots, n_N\rangle = \left(\hat{a}_1^\dagger\right)^{n_1} \left(\hat{a}_2^\dagger\right)^{n_2} \cdots \left(\hat{a}_N^\dagger\right)^{n_N} |0\rangle$ with $|0\rangle = \bigotimes_{i=1}^N |0\rangle_i$ (note that $n_i! = 1$ for the fermion). The operator in Eq. (7) can move its location on this state without changing the sign of the number state: $\hat{v}_i |n_1, n_2, \ldots, n_N\rangle = \left(\hat{a}_1^\dagger\right)^{n_1} \left(\hat{a}_2^\dagger\right)^{n_2} \cdots \left(\hat{a}_{i-1}^\dagger\right)^{n_{i-1}} \hat{v}_i \left(\hat{a}_i^\dagger\right)^{n_i} \cdots \left(\hat{a}_N^\dagger\right)^{n_N} |0\rangle$.

Now, the interaction Hamiltonian considered here reads

$$\hat{H}_I = g\left(\hat{V}^\dagger \hat{V} + N \sum_{i=1}^N |0\rangle_{i\,i}\langle 0|\right), \tag{9}$$

where

$$\hat{V} = \sum_{i=1}^N \hat{v}_i, \tag{10}$$

where g is a coupling constant. The second term inside the brackets on the right-hand side is associated with the subtraction term in Eq. (7), which comes from the property of the phase operator in Eq. (3). Combining Eqs. (1) and (9), we obtain the total Hamiltonian

$$\hat{H} = \hat{H}_0 + \hat{H}_I, \tag{11}$$

which contains both the number and phase operators.

The stationary Schrödinger equation is given by

$$\hat{H}\,|u_E\rangle = E\,|u_E\rangle\,. \tag{12}$$

The solution of this equation is found to be

$$E_{M,N} = M\varepsilon + gf(N), \tag{13}$$

$$\begin{aligned} |u_E\rangle &= |M; N, [\theta_m]\rangle \\ &\equiv \frac{1}{\sqrt{W(M,N)}} \sum_{P\{n\}} |n_1, n_2, \ldots, n_N\rangle\, \delta_{n_1+n_2+\cdots+n_N,M} \exp\left(\mathrm{i}\sum_{i=1}^N n_i\theta_{m_i}\right). \end{aligned} \tag{14}$$

The sum in Eq. (14) implies to be taken over all independent permutations of $(n_1, n_2, \ldots, n_N)$ with the total number of excitations being fixed to be M. Accordingly, quantum entanglement of the specific type is induced by the interaction Hamiltonian. A straightforward calculation shows that $f(N) = N^2$ for the bosons, whereas $0 < f(N) \leq N^2$ for the fermions, depending on the value of M (the maximum N^2 is realized if $M = 1$). $W(M,N)$ is the normalization factor given by

$$W(M,N) = \frac{(M+N-1)!}{(N-1)!M!} \text{ (bosons)}, \quad \frac{N!}{(N-M)!M!} \text{ (fermions)}, \tag{15}$$

which are in fact the degeneracies in Bose-Einstein and Fermi-Dirac statistics. Upon ascertaining that Eq. (14) is in fact the solution, the factor $q^{\hat{F}_i}$ in Eq. (7) is seen to be essential for the fermions.

It is noted that by the gauge transformation in Eq. (4), the operator in Eq. (7) changes as follows: $\hat{v}_i \to \hat{v}_i \exp(\mathrm{i}\theta_{\mu_i})$. Accordingly, the interaction Hamiltonian in Eq. (9) also changes. However, since the state in Eq. (14) transforms as

$$|M; N, [\theta_m]\rangle \to |M; N, [\theta_m + \theta_\mu]\rangle \,, \tag{16}$$

Eq. (12) holds for the gauge transformed Hamitonian and eigenstate with the invariant energy eigenvalues.

Now, Eq. (16) means that the phases in the eigenstate is *indeterminate*. In the vanishing-interaction limit

$$g \to 0, \tag{17}$$

which has been mentioned earlier, the phase operators disappear from the theory. This leads to a situation somewhat analogous to random phases, and therefore the phases are eliminated from the density matrix

$$\begin{aligned} \hat{\rho} &= \frac{1}{(s+1)^N} \sum_{m_1,m_2,\ldots,m_N=0}^{s} |M; N, [\theta_m]\rangle \langle M; N, [\theta_m]| \\ &= \frac{1}{W(M,N)} \sum_{P\{n\}} |n_1, n_2, \ldots, n_N\rangle \langle n_1, n_2, \ldots, n_N| \, \delta_{n_1+n_2+\cdots+n_N, M} . \end{aligned} \tag{18}$$

This is precisely the density matrix in microcanonical ensemble theory, and perfect decoherence in the energy eigenbasis and the equal probability are simultaneously realized, as desired.

Canonical ensemble can be obtained from microcanonical ensemble in a straightforward way. The total isolated system is divided into the objective subsystem and the environmental system, provided that the latter should be much larger than the former. Then, the partial trace is taken over the environmental system to obtain the canonical density matrix (for details, see Ref. [5]).

Thus, we have seen how Bose-Einstein and Fermi-Dirac statistics can be derived from quantum mechanics. From them, classical statistical mechanics simply appears from these in the high-temperature regime, as widely known.

Acknowledgments

YY would like to thank Huaqiao University for hospitality extended to him. The work of SA has been supported in part by grants from National Natural Science Foundation of China (No. 11775084) and Grant-in-Aid for Scientific Research from the Japan Society for the Promotion of Science (No. 26400391 and No. 16K05484), and by the Program of Competitive Growth of Kazan Federal University from the Ministry of Education and Science of the Russian Federation.

References

[1] Goldstein S, Lebowitz J L, Tumulka R and Zanghì N 2006 *Phys. Rev. Lett.* **96** 050403
[2] Reimann P 2007 *Phys. Rev. Lett.* **99** 160404
[3] Gogolin C and Eisert J 2016 *Rep. Prog. Phys.* **79** 056001
[4] D'Alessio L, Kafri Y, Polkovnikov A and Rigol M 2016 *Adv. Phys.* **65** 239

[5] Yokoi Y and Abe S 2018 *J. Stat. Mech.* **2018** 023112
[6] Kobayashi T 1995 *Phys. Lett.* A **207** 320
[7] Kobayashi T 1998 *Nuovo Cimento* B **113** 633
[8] Abe S and Kobayashi T 2003 *Phys. Rev.* E **67** 036119
[9] Pegg D T and Barnett S M 1988 *Europhys. Lett.* **6** 483
[10] Bendjaballah C 1995 *Introduction to Photon Communication* (Berlin: Springer)
[11] Dirac P A M 1927 *Proc. R. Soc. London Ser.* A **114** 243
[12] Fujikawa K 1995 *Phys. Rev.* A **52** 3299

Composition of Fractals

Y. Lanoiselée[1,2,3], **L. Nivanen**[3], **A. El Kaabouchi**[3], **Q. A. Wang**[4,5]

[1]Laboratoire de Physique de la Matiére Condensée, Ecole Polytechnique, CNRS, 91128 Palaiseau Cedex, France.
[2]Polytech'Tours, 64 Avenue Jean Portalis, 37200 Tours, France.
[3]Laboratoire de Physique Statistique et Systèmes Complexes, ISMANS, 44 Avenue, F.A. Bartholdi, 72000 Le Mans, France.
[4]ESIEA GROUP, 9 rue Vesale, 75005 Paris, France
[5]IMMM, UMR 6283, Université du Maine, Avenue O. Messiaen, 72085 Le Mans Cedex, France.

E-mail: alexandre.wang@univ-lemans.fr

Abstract. This work is an analytical and numerical study of the composition of several fractals into one and of the relation between the composite dimension and the dimensions of the component fractals. In the case of composition of standard IFS with segments of equal size, the composite dimension can be expressed as a function of the component dimensions. But in the case of the compositions including component multifractals, the composite dimension cannot be expressed as explicit function of component dimensions and can only be solved numerically. An application of fractal composition to a physics problem within the incomplete statistics is discussed.

1. Introduction

A fractal is a geometrical set that typically displays self-similar patterns which may be exactly or nearly the same at different scales [1, 2, 3, 4]. In general, the concept of fractal extends beyond self-similarity and includes the case of patterns that display details at every scale. A fractal scales differently from the ordinary geometrical figures in that its box dimension α [3] is not necessarily integer. For instance, an ordinary line has $\alpha = 1$, and an ordinary area has $\alpha = 2$. But for a fractal line α can be more or less than unity. The fractal dimension is not the unique parameter describing a fractal structure, but it is the most important one characterizing the scaling property of fractals and has played important role in the application of fractal geometry to other scientific and technological domains as a modeling and design tool [1, 2, 3].

In this work, we focus on the mathematical operation of composing different fractals or multifractals and the calculation of the fractal dimension. By composition of fractals, we mean an operation to mix two or more fractals (multifractals) in some manner to construct a single composite fractal or multifractal. The mixing of different fractals can be performed in different manners by alternating the iterations of the component fractals in a deterministic or probabilistic way. A probabilistic mixing of different fractals has been recently proposed by Barnsley et al in the framework of the V-fractals and superfractals [5, 6] in order to facilitate the modeling of natural phenomena by superfractals. The present work is focused on the deterministic mixing of fractals. The main aim is to find the relationship between the dimensions of the component fractals and the dimension of the composite fractal, and to see whether or not this

IOP Conf. Series: Journal of Physics: Conf. Series **1113** (2018) 012013 doi:10.1088/1742-6596/1113/1/012013

relationship is dependent on the mixing manner. This question has been raised several years ago within a statistical theory called incomplete statistics (IS) which has been proposed by physical consideration in order to make statistics in, amount others, fractal phase spaces of dynamic systems [8, 9, 10, 11, 12, 13]. A problem in this framework was to find the characteristic parameter (associated to the fractal dimension, see section 5.2 below) of a composite system from the same parameters of the component systems with the condition of the thermodynamic equilibrium [13, 15, 16]. The second aim of this work is to justify a hypothesis of the IS (see section 5.2 below) under the angle of the composition of fractals.

2. Composition of fractals

A fractal can be obtained from an Iterated Function System (IFS) [3, 17]. An IFS is a mapping in which, starting with a closed subspace of $\mathbb{R}^n$, we iterate infinitely a set of contractions on it. A fractal is generated as the attractor of an IFS in which all contractions have the same scale factor. So, it can be described by two numbers :

- N : the number of copies (contractions) created at each stage from one copy of the previous stage with $N \in \mathbb{N}^*$.
- ρ : the scale factor applied on each copy. We have $0 < \rho < 1$ verified in the following for any IFS case and any generalization.

2.1. Box dimension

We start with the calculation of the box dimension for a fractal : the Koch curve (Fig.1). The construction of this fractal results in an infinite curve length. In what follows, we will use the content $C_{k,\alpha}$ of the Koch curve from satge k. The number $C_{k,\alpha}$ is defined by the $\alpha-$ dimensional Hausdorff $\mathcal{H}^\alpha$ of F_k, where F_k is the curve at the kth iteration [3]. For an ordinary line, $\alpha = 1$ and for an ordinary area, $\alpha = 2$. For a fractal shape, α is not necessarily an integer.

At each stage of the iteration of an IFS, N copies are produced with a scale factor ρ from every previous copy. We begin with an initial segment of length L_0 which is called the initiator. F_k is consisted of $N_\eta = N^k$ copies of length $\eta = \rho^k L_0$.

The content $C_{k,\beta}$ is obtained with : $C_{k,\beta} = N_\eta \eta^\beta$ [3]. We can write : $C_{k,\beta} = (N\rho^\beta)^k L_0^\beta$. A fractal is only defined when $k \to \infty$. To reach a finite content and obtain the real dimension α, it's necessary that α verifies $N\rho^\alpha = 1$ which can be written as

$$\ln N + \alpha \ln \rho = 0$$

or

$$\alpha = \frac{\ln N}{\ln \frac{1}{\rho}} \tag{1}$$

which is the box dimension. In the case of Fig.1, $\alpha = \frac{\ln 4}{\ln 3} \approx 1.26$.

2.2. Composition of 2 fractals

The set of contractions of an IFS which is applied at each iteration is a geometric transformation. When the scale factors of the set of contractions are equals, the transformation can be expressed as a sequence of angles. Koch curve can be written as $\left[0, \frac{\pi}{3}, -\frac{\pi}{3}, 0\right]$. In what follows, we call it $K_{\frac{\pi}{3}}$. A modified Koch curve with angle θ can be written as K_θ. For an IFS, the same transformation is applied until infinity. Koch curve can be represented by a infinite sequence of transformations as follows:

$$Koch\,Curve = K_{\frac{\pi}{3}}\, K_{\frac{\pi}{3}}\, K_{\frac{\pi}{3}}\, K_{\frac{\pi}{3}}\, K_{\frac{\pi}{3}}\, K_{\frac{\pi}{3}}\, K_{\frac{\pi}{3}} ... = \underline{K_{\frac{\pi}{3}}}.$$

IOP Conf. Series: Journal of Physics: Conf. Series **1113** (2018) 012013 doi:10.1088/1742-6596/1113/1/012013

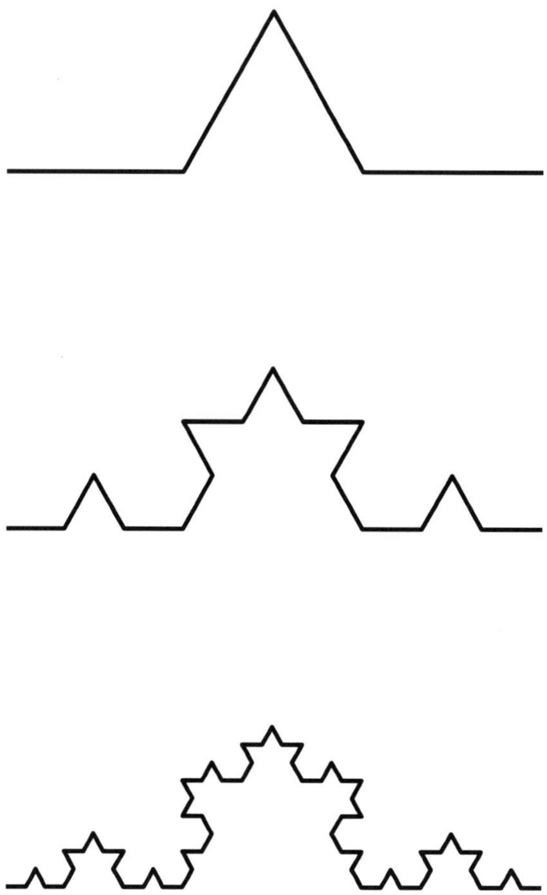

Figure 1. Koch Curve ($N = 4$, $\rho = 1/3$) from stage $k = 1$ to 3. The dimension of this curve is $\alpha = \frac{\ln 4}{\ln 3} \approx 1.26$.

Here we are interested in composing fractals in the following way: at each stage we apply once the transformation associated with the first fractal and once the transformation associated with the second fractal. If we want to compose Koch curve by a modified Koch curve with angle of $\frac{\pi}{4}$, the composite fractal can be represented by

$$Koch\ by\ Modified\ Koch = K_{\frac{\pi}{4}}K_{\frac{\pi}{3}}\ K_{\frac{\pi}{4}}K_{\frac{\pi}{3}}\ K_{\frac{\pi}{4}}K_{\frac{\pi}{3}}... = \underline{K_{\frac{\pi}{4}}K_{\frac{\pi}{3}}}$$

where each transformation is called a substage and a period of transformation is called a stage. Hence each stage is composed of a $K_{\frac{\pi}{4}}$ substage and a $K_{\frac{\pi}{3}}$ substage. The underline means the couple of transformations to be repeated until infinity. Each stage associated with the composite fractal produces respectively N_1 and N_2 copies with respectively scale factors ρ_1 and ρ_2 (Fig.2).

Hence at the first substage, N_1 copies are generated with a scale factor ρ_1. And at the second substage N_2 copies are generated with a scale factor ρ_2 from each of the N_1 copies of the first substage. So at stage k we have $N_\eta = {N_1}^k {N_2}^k$ copies of length $\eta = {\rho_1}^k {\rho_2}^k L_0$. The content is

$$C_{k,\beta} = (N_1 N_2 (\rho_1 \rho_2)^\beta)^k {L_0}^\beta. \tag{2}$$

To have a finite content when $k \to \infty$, it's necessary that : $N_1 N_2 (\rho_1 \rho_2)^\alpha = 1$, which leads to :

$$\alpha = \frac{\ln N_1 + \ln N_2}{\ln \frac{1}{\rho_1} + \ln \frac{1}{\rho_2}}. \tag{3}$$

The dimension of the composite fractal can be expressed as a function of the dimensions of the component fractals. Remember that each component fractal verifies the relation $N_i \rho_i^{\alpha_i} = 1$ with $i \in \{1, 2\}$ where α_i is its dimension, we can write $N_i = \frac{1}{\rho_i^{\alpha_i}}$ and replace it in the above equation to give $\alpha = \frac{\ln \frac{1}{{\rho_1}^{\alpha_1}} + \ln \frac{1}{{\rho_2}^{\alpha_2}}}{\ln \frac{1}{\rho_1} + \ln \frac{1}{\rho_2}}$ and :

$$\alpha = \frac{\alpha_1 \ln \rho_1 + \alpha_2 \ln \rho_2}{\ln \rho_1 + \ln \rho_2}. \tag{4}$$

The composite dimension is independent of the order of the substages in the IFS. Note that its expression is in the form of a barycentric average with positive coefficients $\frac{\ln \rho_1}{\ln \rho_1 + \ln \rho_2}$ and $\frac{\ln \rho_2}{\ln \rho_1 + \ln \rho_2}$. Because of this average, we know that α lies between α_1 and α_2. For Fig.2, using Eq.(3), we can calculate $\alpha \approx 1.19$.

2.3. Composition of m fractals

This section describes how to obtain the dimension of a composite fractal composed of m different fractals. We construct it by an IFS of m substages (corresponding to m sets of contractions) in each stage of the iteration. Let T_i be the transformation associated with the ith fractal, we can represent the composition with:

$$Composition\ of\ m\ fractals = T_1T_2...T_mT_1T_2...T_mT_1T_2...T_m... = \underline{T_1T_2...T_m}.$$

2.3.1. General Case We make m substages with N_i copies of scale factor ρ_i for the substage i with $i \in [[1, m]]$. Each substage verifies $N_i {\rho_i}^{\alpha_i} = 1$ with α_i the dimension of the ith component fractal. Fig.3 shows an example combining a Cantor set, a quadratic Koch curve (noted $Q_{\frac{\pi}{2}}$) and a Koch curve.

At the kth stage, $N_\eta = \prod_{i=1}^m N_i^k$ copies of length $\eta = \left(\prod_{i=1}^m \rho_i^k\right) L_0$ are produced. The content is:

$$C_{k,\beta} = (\prod_{i=1}^{m} N_i {\rho_i}^\beta)^k {L_0}^\beta. \tag{5}$$

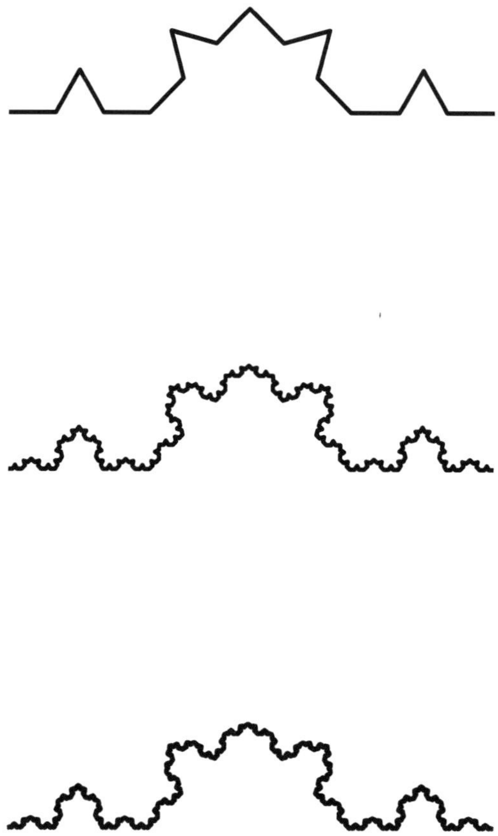

Figure 2. A modified Koch curve with angles of $\frac{\pi}{4}$ ($N_1 = 4$, $\rho_1 = \frac{1}{2+\sqrt{2}}$) composed with a Koch Curve ($N_2 = 4$, $\rho_2 = 1/3$). It's the result for $\underline{K_{\frac{\pi}{4}} K_{\frac{\pi}{3}}}$ for stage $k = 1$ to 3. The component dimension of the curve is $\alpha \approx 1.19$.

When $k \to \infty$, it's necessary that :

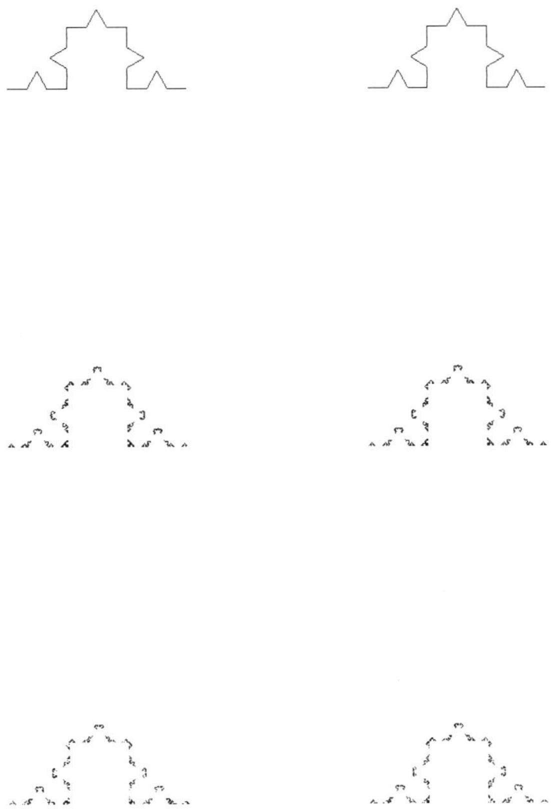

Figure 3. A composite IFS composed of a Cantor set ($N_1 = 2$, $\rho_1 = 1/3$), a quadratic Koch curve ($N_2 = 5$, $\rho_2 = 1/3$) and a Koch curve ($N_3 = 4$, $\rho_3 = 1/3$). The composition rule is $Composition = \underline{C_{[1/3\ 1/3]} Q_{\frac{\pi}{2}} K_{\frac{\pi}{3}}}$ for the stage k with $k = 1$ to 3. The composite dimension is $\alpha \approx 1.12$.

$$\prod_{i=1}^{m} N_i {\rho_i}^{\alpha} = 1, \tag{6}$$

or

$$\alpha = \frac{\sum_{i=1}^{m} \ln N_i}{\sum_{i=1}^{m} \ln \frac{1}{\rho_i}}.$$

For the example in Fig.3, from Eq.(7) : $\alpha =\approx 1.12$.

Because of the commutativity of the sum in Eq.(7). The dimension α is independent of the order of application of different transformations T_i, with $i \in [[1, m]]$ in a stage. The Eq.(7) is a generalization of the result obtained in Eq.(3).

To express the composite dimension as a function of the component dimensions and scale factors, we use $N_i {\rho_i}^{\alpha_i} = 1$ to replace N_i in Eq.(7) and obtain

$$\alpha = \frac{\sum_{i=1}^{m} \alpha_i \ln \rho_i}{\sum_{i=1}^{m} \ln \rho_i}.$$

If the m component IFS have the same scale factor, we can write $\rho_i = \rho$.Eq.(7) becomes

$$\alpha = \frac{\sum_{i=1}^{m} \alpha_i}{m}$$

which is an arithmetic average of the component dimensions.

We can also express the composite dimension as a function of the component dimensions and the number of copies of each substage N_i. Let us take Eq.(7) and replace ρ_i by $\rho_i = \frac{1}{N_i^{\frac{1}{\alpha_i}}}$. The composite dimension then verifies

$$\frac{1}{\alpha} = \frac{\sum_{i=1}^{m} \frac{1}{\alpha_i} \ln N_i}{\sum_{i=1}^{m} \ln N_i}$$

where the inverse composite dimension is in the form of a barycentric average of the component dimension determined by the integer numbers N_i.

If the m component fractals have the same number of copies, we can write $N_i = N$ and obtain

$$\frac{1}{\alpha} = \frac{\sum_{i=1}^{m} \frac{1}{\alpha_i}}{m}$$

i.e., the composed dimension is the harmonic average of the component dimensions.

3. Composition of multifractals

A multifractal set is the attractor of an IFS where the scale factors can be different. At a given stage, the segments have different sizes and forms which can be treated with multinomial expansions. Let r_j be the multifractal scale factors with $j \in [[1, l]]$ where l is the number of scaling applied. As in the case of fractals, we shall verify $\forall j \in [[1, l]]$, $0 < r_j < 1$ for any multifractal. To have non zero finite content, each multifractal verifies the Moran equation[13]:

$$\sum_{j=1}^{l} {r_j}^{\alpha} = 1$$

with α the dimension of the multifractal. If all scale factors are equal $r_j = r$, the Moran equation becomes $lr^\alpha = 1$, the relation for standard IFS.

It is possible to prove the uniqueness of α. We take a function $f(\alpha) = \sum_{j=1}^{l} {r_j}^{\alpha}$. It is a continuous and constantly decreasing function. When $\alpha \to 0$, $f(\alpha) \to l$ and when $\alpha \to \infty$, $f(\alpha) \to 0$. So $\sum_{j=1}^{l} {r_j}^{\alpha} = 1$ has unique solution corresponding to the box dimension of the curve.

3.1. Composition of a fractal and a binary multifractal

A binary multifractal is constructed with two different scale factors r_1 and r_2, with $\{r_1, r_2\} < 1$ and ${r_1}^{\alpha_r} + {r_2}^{\alpha_r} = 1$ [13], where α_r is the dimension of the binary multifractal. In what follows, we focus on the composition of a binary multifractal with a fractal. Fig.4 gives an example of composition of a binary multifractal and a Koch curve.

Each stage of the iteration is composed of two substages: one being the iteration of the binary multifractal and another one is the iteration of the fractal. At the kth stage F_k is composed of $N_k = 2^k N^k$ copies. At a given scale, the length of a segment is $\delta_{k,p}$ with $p \in [[1, N_k]]$, so the total length of the curve is $L_k = \sum_{p=1}^{N_k} \delta_{k,p}$. Now consider $j \in [[0, k]]$, we can find $\binom{k}{j}$ segments of length $\Delta_{k,j} = {r_1}^j {r_2}^{k-j} L_0$ on the binary multifractal[13]. Each segment of this length contains N^k segments of length $\rho^k L_0$ generated by fractal. It is obvious that at the kth stage we have $\binom{k}{j} N^k$ segments of length $\Delta_{k,j} = {r_1}^j {r_2}^{k-j} \rho^k L_0$. The total length of the curve is then:

$$L_k = \sum_{p=1}^{N_k} \delta_{k,p} = \sum_{j=0}^{k} \binom{k}{j} N^k {r_1}^j {r_2}^{k-j} \rho^k L_0 = N^k \rho^k \left(r_1 + r_2\right)^k L_0.$$

The content of this form is given by:

$$C_{k,\beta} = \sum_{p=1}^{N_k} {\delta_{k,p}}^{\beta} = \sum_{j=0}^{k} \binom{k}{j} {r_1}^{\beta j} {r_2}^{\beta(k-j)} N^k \rho^{\beta k} {L_0}^{\beta} \tag{7}$$

in which the binomial expression is

$$\sum_{j=0}^{k} \binom{k}{j} r_1^{\beta j} r_2^{\beta(k-j)} = (r_1^\beta + r_2^\beta)^k.$$

Hence the content of order β reads:

$$C_{k,\beta} = \left[({r_1}^{\beta} + {r_2}^{\beta}) N \rho^{\beta}\right]^k {L_0}^{\beta}. \tag{8}$$

The content must be finite when $k \to \infty$, implying

$$({r_1}^{\alpha} + {r_2}^{\alpha}) N \rho^{\alpha} = 1. \tag{9}$$

This expression gives a unique α as the composite dimension of the curve. The dependence of α on the component dimension α_ρ of the fractal can be shown as follows. α_ρ verifying $N\rho^{\alpha_\rho} = 1 \Leftrightarrow N = \frac{1}{\rho^{\alpha_\rho}}$, the above equation can be written as

$$({r_1}^{\alpha} + {r_2}^{\alpha}) \rho^{\alpha - \alpha_\rho} = 1. \tag{10}$$

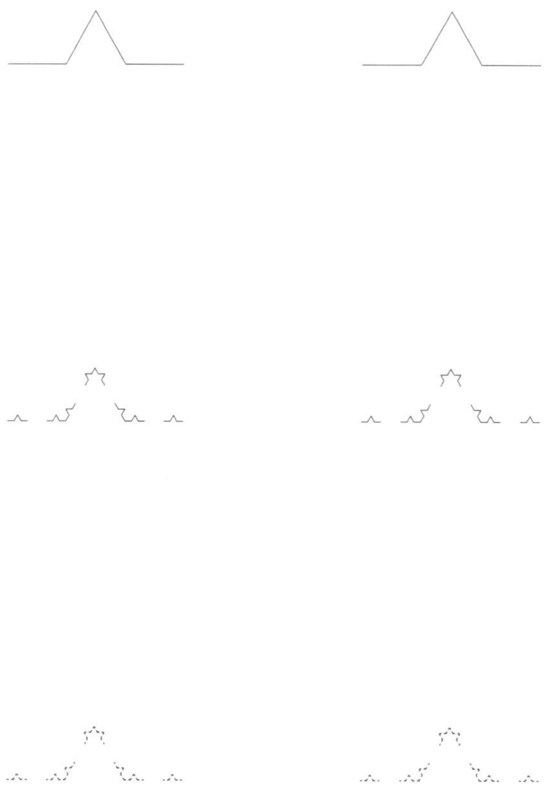

Figure 4. Composition of a binary multifractal ($r_1 = 1/2$, $r_2 = 1/3$) with a Koch Curve ($N = 4$, $\rho = 1/3$) The composition rule is $Composition = \underline{C_{[1/2\ 1/3]} K_{\frac{\pi}{3}}}$ for the stage k with $k = 1$ to 3.

However, in general it is impossible to introduce the multifractal dimension α_r. α should be obtained by numerical solution of the above two equations.

Anyway, analytical solution of α from Eq.(10) is possible under special conditions. For example, one of these conditions is $r_2 = r_1^2\rho$. The proof is the following.

Let us put r_2 into Eq.(10) to get

$$({r_1}^{\alpha} + {r_1}^{2\alpha}\rho^{\alpha})\rho^{\alpha-\alpha_\rho} = 1$$

or

$$({r_1}^{\alpha}\rho^{\alpha})^2 + {r_1}^{\alpha}\rho^{\alpha} - \rho^{\alpha_\rho} = 0$$

.

Because $(r_1\rho)^\alpha > 0$ and $\rho^{\alpha_\rho} = \frac{1}{N}$ it follows that

$$(r_1\rho)^\alpha = \frac{-1 + \sqrt{1 + \frac{4}{N}}}{2}$$

Finally:

$$\alpha = \frac{\ln \frac{-1+\sqrt{1+\frac{4}{N}}}{2}}{\ln r_1\rho}. \tag{11}$$

Fig.5 shows an example of composition of a Cantor ($r_1 = 1/2$, $r_2 = 1/12$) set with a Koch curve ($N = 4$, $\rho = 1/3$).

where $r_2 = \left(\frac{1}{2}\right)^2 \frac{1}{3} = \frac{1}{12}$.

The dimension is given by $\alpha = \frac{\ln \frac{-1+\sqrt{1+\frac{4}{4}}}{2}}{\ln(\frac{1}{2}\frac{1}{3})} \approx 0.88$.

3.2. Composition of m fractals and a multifractal

Now we focus on the composition of m different fractals and a multifractal. At the kth stage, each one of the m transformations associated with fractals produce N_i copies of scale factor ρ_i with $i \in [[1, m]]$, with $\prod_{i=1}^{m} {N_i}^k$ the total number of segments of scale factor $\prod_{i=1}^{m} {\rho_i}^k$. On the other hand, the multifractal set is defined by l different scale factors r_j with $r_j < 1$ and $j \in [[1, l]]$. An example of this composition is given in Fig.6.

At the kth stage, on F_k we find $N_k = l^k \prod_{i=1}^{m} {N_i}^k$ segments of length $\delta_{k,p}$ with $p \in [[1, N_k]]$. Let G be any set of integers $\{g_1, g_2, ..., g_m\}$ verifying :

$$\forall j \in [[1, l]],\ g_j \in [[0, k]],\ \sum_{j=1}^{l} g_j = k. \tag{12}$$

For any set G that verifies Eq.(12), we can find, on a multifractal, $\frac{k!}{\prod_{j=1}^{l} g_j!}$ segments of length

$\Delta_{k,G} = (\prod_{j=1}^{l} {r_j}^{g_j})L_0$ [13]

At the kth stage, on the F_k form, we have $\frac{k!}{\prod_{j=1}^{l} g_j!} \prod_{i=1}^{m} {N_i}^k$ segments of length : $\Delta_{k,G} = \left(\prod_{j=1}^{l} {r_j}^{g_j}\right)\left(\prod_{i=1}^{m} {\rho_i}^k\right) L_0$.

Then the total length is :

Figure 5. A binary multifractal ($r_1 = 1/2$, $r_2 = 1/12$) composed with Koch curve ($N = 4$, $\rho = 1/3$, $\alpha_\rho \approx 1.26$). It's the result for $C_{[1/2\ 1/12]}K_{\frac{\pi}{3}}$ for stage $k = 1$ to 3. The component dimension of the curve is $\alpha \approx 0.88$.

$$
\begin{aligned}
L_k &= \sum_{p=1}^{N_k} \delta_{k,p} \\
&= \sum_{G} \left(\frac{k!}{\prod_{j=1}^{l} g_j!} \right) \left(\prod_{j=1}^{l} r_j^{\,g_j} \right) \left(\prod_{i=1}^{m} N_i \rho_i \right)^k L_0
\end{aligned}
$$

$$
\left(\sum^{l} \quad \right)^k \left(\prod^{m} \quad \right)^k
$$

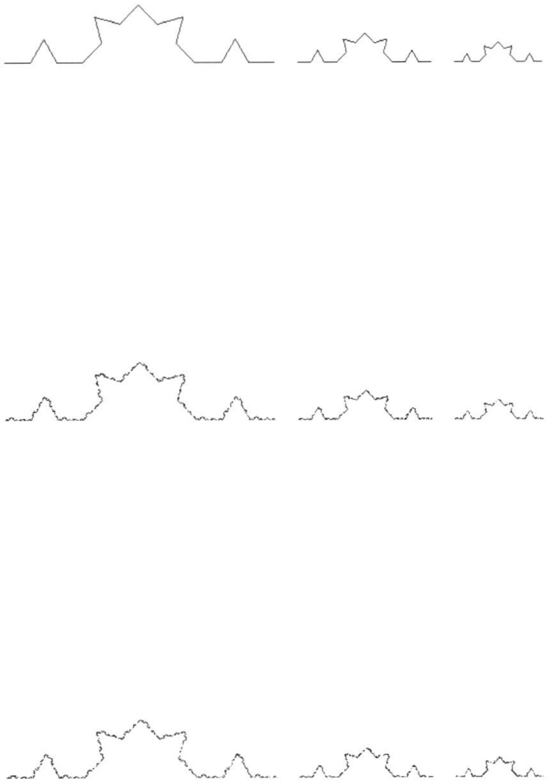

Figure 6. Composition, from stage 1 to 3, of a multifractal ($r_1 = 1/2$, $r_2 = 1/4$, $r_3 = 1/6$), a modified Koch curve for $\theta = \pi/4$ ($N_1 = 4$, $\rho_1 = \frac{1}{2+\sqrt{2}}$) and a Koch curve ($N_2 = 4$, $\rho_2 = 1/3$). This composition is represented by :
$Composition = C_{[1/2,1/4,1/6]} K_{\frac{\pi}{4}} K_{\frac{\pi}{3}} C_{[1/2,1/4,1/6]} K_{\frac{\pi}{4}} K_{\frac{\pi}{3}} ... = \underline{C_{[1/2,1/4,1/6]} K_{\frac{\pi}{4}} K_{\frac{\pi}{3}}}.$

Now let us calculate the content of order α on F_k :

$$C_{k,\beta} = \sum_{p=1}^{N_k} {\delta_{k,p}}^{\beta} = \sum_{G} \left(\frac{k!}{\prod\limits_{j=1}^{l} g_j!} \right) \left(\prod_{j=1}^{l} {r_j}^{\beta g_j} \right) \left(\prod_{i=1}^{m} N_i {\rho_i}^{\beta} \right)^k L_0^{\beta}.$$

We have here a multinomial expansion.

$$\sum_{G}\left(\frac{k!}{\prod_{j=1}^{l} g_j!}\right)\left(\prod_{j=1}^{l} {r_j}^{\beta g_j}\right) = \left(\sum_{j=1}^{l} {r_j}^{\beta}\right)^{k}. \tag{14}$$

It leads to :

$$C_{k,\beta} = \left[\left(\prod_{i=1}^{m} N_i {\rho_i}^{\beta}\right)\left(\sum_{j=1}^{l} {r_j}^{\beta}\right)\right]^{k} {L_0}^{\beta}. \tag{15}$$

When $k \to \infty$ the content remains finite with the dimension verifying :

$$\left(\prod_{i=1}^{m} N_i {\rho_i}^{\alpha}\right)\left(\sum_{j=1}^{l} {r_j}^{\alpha}\right) = 1 \tag{16}$$

or

$$\left(\prod_{i=1}^{m} {\rho_i}^{\alpha-\alpha_{\rho,i}}\right)\left(\sum_{j=1}^{l} {r_j}^{\alpha}\right) = 1. \tag{17}$$

3.3. Composition of m multifractals

3.3.1. Composite dimension In this section we study the composition of m different multifractals. Each one of these multifractals is defined by l_i scale factors with $i \in [[1, m]]$ corresponding to the sets of scale factors from each one of the m multifractals.

At the kth stage, F_k is composed of $N_k = \prod_{i=1}^{m} {l_i}^{k}$ segments of length $\delta_{k,p}$ with $p \in [[1, N_k]]$.

We note G_i any set of integers $\{g_{i,1}, g_{i,2}, \cdots, g_{i,l_i}\}$ that verifies:

$$\forall j \in \{1 \cdots l_i\},\ g_{i,j} \in \{0, 1, \cdots, k\},\ \sum_{j=1}^{l_i} g_{i,j} = k. \tag{18}$$

For any set G_i that verifies Eq.(18) :

We can find $\frac{k!}{\prod_{j=1}^{l_i} g_{i,j}!}$ segments of length $\Delta_{k,G_i} = (\prod_{j=1}^{l_i} {r_{i,j}}^{g_{i,j}})L_0$.

The total length of the F_k form is given by:

$$L_k = \sum_{p=1}^{N_k} \delta_{k,p} = \prod_{i=1}^{m}\left[\sum_{G_i}\left(\frac{k!}{\prod_{j=1}^{l_i} g_{i,j}!}\right)\left(\prod_{j=1}^{l_i} {r_{i,j}}^{g_{i,j}}\right)\right] L_0 = \prod_{i=1}^{m}\left(\sum_{j=1}^{l_i} r_{i,j}\right)^{k} L_0.$$

The content of order β of the form F_k is:

$$C_{k,\beta} = \sum_{p=1}^{N_k} {\delta_{k,p}}^{\beta} = \prod_{i=1}^{m}\left[\sum_{G_i}\left(\frac{k!}{\prod_{j=1}^{l_i} g_{i,j}!}\right)\left(\prod_{j=1}^{l_i} {r_{i,j}}^{\beta g_{i,j}}\right)\right] {L_0}^{\beta}$$

This is the form of a product of multinomial expansions.

$$\prod_{i=1}^{m}\left[\sum_{G_i}\left(\frac{k!}{\prod_{j=1}^{l_i} g_{i,j}!}\right)\left(\prod_{j=1}^{l_i} {r_{i,j}}^{\beta g_{i,j}}\right)\right]=\prod_{i=1}^{m}\left(\sum_{j=1}^{l_i} {r_{i,j}}^{\beta}\right)^{k}$$

. The content is recast as:

$$C_{k,\beta}=\prod_{i=1}^{m}\left(\sum_{j=1}^{l_i} {r_{i,j}}^{\beta}\right)^{k} {L_0}^{\beta}$$

.

Because the content is finite when $k \to \infty$, we have

$$\prod_{i=1}^{m}\left(\sum_{j=1}^{l_i} {r_{i,j}}^{\alpha}\right)=1. \tag{19}$$

As all $r_{i,j} < 1$, the expression is strictly decreasing so there is a unique α which satisfies Eq.(19). Since multifractal is a general case of fractals, Eq.(19) is a general formula of composite dimension for all compositions.

3.3.2. Borders of the composite dimension Since the composition of m multifractals is a generalization of the composition of m fractals, it is reasonable to think of the existence of borders for the composite dimension such as $\underset{i\in[[1,m]]}{min}(\alpha_i) \le \alpha \le \underset{i\in[[1,m]]}{max}(\alpha_i)$. This can be proved by contradiction. Let's postulate $\underset{i\in[[1,m]]}{max}(\alpha_i) < \alpha$. The m multifractals verify $\sum_{j=1}^{l_i} {r_{i,j}}^{\alpha_i} = 1$ with $i \in [[1,m]]$. Let $f_i(\alpha_i)$ be a function of the component dimensions α_i such that $f_i(\alpha_i) = \sum_{j=1}^{l_i} {r_{i,j}}^{\alpha_i}$. Then all f_i are strictly decreasing functions so:

$$\forall i \in [[1,m]],\ f_i(\alpha) < f_i(\alpha_i) = 1.$$

Hence

$$\prod_{i=1}^{m} f_i(\alpha) < \prod_{i=1}^{m} f_i(\alpha_i) = 1.$$

But from Eq.(19), we know that $\prod_{i=1}^{m} f_i(\alpha) = 1$. This is a contradiction, hence we must have

$$\underset{i\in[[1,m]]}{min}(\alpha_i) \le \alpha \le \underset{i\in[[1,m]]}{max}(\alpha_i). \tag{20}$$

We can conclude that the composite dimension is always between the minimum and the maximum of component dimensions for any composition.

4. Periodically alternating compositions

The results presented above concerns the compositions performed by alternating the iterations of component fractals one after another, meaning that in each stage of the composite iteration, the number of substages is equal to the number of different component fractals. In what follows, we will study a more general case where, in a stage of the composite iteration, each component fractal is repeated several times following by repeating the iteration of another component fractals.

4.1. For 2 fractals

Consider two component fractals whose corresponding transformations produces in each iteration N_1 and N_2 copies of scale factor ρ_1 and ρ_2, respectively. At each stage of the composition, we apply n_1 times the first scaling and n_2 times the second scaling. So each stage is composed of $n = n_1 + n_2$ substages. In the following Figure, we apply once Cantor and twice Koch transformation in each stage as represented by the following sequence:

$$Cantor_{[1/3,1/3]}\ by\ Koch_{\frac{\pi}{3}} = CKK\ CKK\ CKK\ CKK\ CKK... = \underline{CKK}.$$

We can use Eq.(7) to find the composite dimension as if we had $n = n_1+n_2$ different substages (component fractals):

$$\alpha = \frac{\sum_{i=1}^{n} \alpha_i \ln \rho_i}{\sum_{i=1}^{n} \ln \rho_i}.$$

It is clear that in each stage of n substages, there are n_1 times the first scaling and n_2 times the second one. Hence we have

$$\alpha = \frac{n_1\alpha_1 \ln \rho_1 + n_2\alpha_2 \ln \rho_2}{n_1 \ln \rho_1 + n_2 \ln \rho_2}. \tag{21}$$

Hence we conclude that the composite dimension is independent of the order of substages in a stage if n_1 and n_2 are given. Consequently, the example in Fig.7 has the same dimension as the following sequence:

$$Cantor_{[1/3,1/3]}\ by\ Koch_{\frac{\pi}{3}} = KCK\ KCK\ KCK\ KCK\ KCK\cdots = \underline{KCK}$$

Further more, the order of substages can be different from one stage to another without affecting the composite dimension if n_1 and n_2 are given. Hence the following sequence has the same composite dimension as in Fig.7.

$$Cantor_{[1/3,1/3]}\ by\ Koch_{\frac{\pi}{3}} = KCK\ CKK\ KKC\ CKK\ KCK\cdots.$$

4.2. For m fractals

In this part we will extend the case of a periodic composition of 2 fractals to m fractals, by applying n_i times one of the m transformation to produce N_i segments of scale factor ρ_i. For example we can make a composition of 3 fractals whose iterations are represented by T_1, T_2 and T_3, respectively. T_1 is repeated $n_1 = 3$ times, T_2 $n_2 = 2$ times and T_3 $n_3 = 1$ times. Hence there are 6 substages in each stage of the composition. The composition can be represented by the following sequence:

$$Composition = T_1T_1T_1T_2T_2T_3\ T_1T_1T_1T_2T_2T_3\ldots = \underline{T_1T_1T_1T_2T_2T_3}.$$

So in general each stage is composed of $\sum_{i=1}^{m} n_i = n$ substages. By using Eq.(7), we obtain

$$\alpha = \frac{\sum_{i=1}^{n} \alpha_i \ln \rho_i}{\sum_{i=1}^{n} \ln \rho_i} = \frac{\sum_{i=1}^{m} n_i\alpha_i \ln \rho_i}{\sum_{i=1}^{m} n_i \ln \rho_i}.$$

If all n_i are the same, we find Eq.(7). It is obvious that the composite dimension, for given fractal, depends on the number of repetition of each IFS in each stage of the composition.

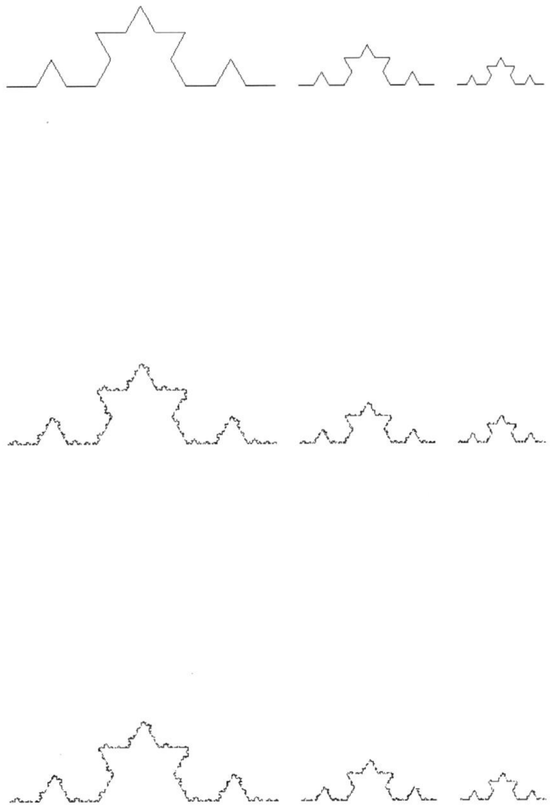

Figure 7. Cantor applied once and Koch curve applied twice by stage,that makes 3 substages, with respectively ($n_1 = 1$, $N_1 = 2$, $\rho_1 = 1/3$) and ($n_2 = 2$, $N_2 = 4$, $\rho_2 = 1/3$) from stage $k = 1$ to 3. The dimension obtained from Eq.(21) is $\alpha \approx 1.05$.

4.3. For m multifractals

We can also make, in the same way as above, periodic composition for multifractals. Each one of the m multifractal transformation is applied n_i times in each stage of the composition. The total number of substages is then $\sum_{i=1}^{m} n_i = n$. For example, 3 multifractals, M_1, M_2 and M_3, are composed periodically. Suppose $n_1 = 1$, $n_2 = 2$ and $n_3 = 3$. There are 6 substages in each stage of the composition. The sequence is :

$$\begin{aligned} Composition &= M_1M_2M_2M_3M_3M_3\ M_1M_2M_2M_3M_3M_3\ ... \\ &= \underline{M_1M_2M_2M_3M_3M_3} \end{aligned} \tag{22}$$

The general formula for this composition is:

$$\prod_{i=1}^{m} \left(\sum_{j=1}^{l_i} r_{i,j}{}^{\alpha} \right)^{n_i} = 1.$$

Because the relation is commutative, the composite dimension does not depend on the order of the different substages. It depends only on the numbers n_i.

5. Application

5.1. How to get a dimension close to a rational number

Starting with an initial fractal of any dimension, we can get a composite fractal with dimension as close as we want to a rational number. Consider a fractal constructed by a transformation which creates N_1 copies at each iteration with a scale factor ρ_1. It is composed with another fractal whose transformation creates $N_2 = n^{a_1}$ copies with of scale factor $\rho_2 = \frac{1}{n^{a_2}}$ where $\{n, a_1, a_2\} \in \mathbb{N}^*$. The composite dimension α is given by

$$\alpha_{lim} = \lim_{n\to\infty} \frac{\ln N_1 + \ln n^{a_1}}{\ln \frac{1}{\rho_1} + \ln n^{a_2}} = \lim_{n\to\infty} \frac{\frac{\ln N_1}{\ln n^{a_1}} + 1}{\frac{\ln \frac{1}{\rho_1}}{\ln n^{a_1}} + \frac{\ln n^{a_2}}{\ln n^{a_1}}} = \frac{a_1}{a_2}.$$

So it is possible to get a dimension as near as possible (depending on n) to a rational number $\frac{a_1}{a_2}$.

5.2. Connection with Incomplete Statistics

We study here the link between fractal composition and the incomplete statistics [8] which has been proposed several years ago in order to make statistical study of physical systems for which the mathematical calculation of available states is incomplete (under or over counted). One of the possible applications is to the cases where the phase space is fractal or the probability measure can be represented in fractal geometry. For instance, for a multifractal phase space, the incomplete probability distribution is defined for any segment i_k at any stage k by $p_{i_k} = \delta_{i_k}/L_0$ with the normalization $\sum_{i=1}^{l} p_{i_k}^{\alpha} = 1$ where α is a ratio equal to the box dimension d_f divided by the dimension d of the space into which the fractal is embedded: $\alpha = d_f/d$[13]. α characterizes the incompleteness of the statistics on that fractal phase space. As this incomplete normalization is valid for any k, we will drop the index k from the normalization in what follows.

The incomplete statistics has been proposed in the context of the study of a nonextensive statistics[8, 9]. In an effort to apply the nonextensive statistics to thermodynamic systems composed of two independent subsystems, a problem is encountered about the derivation of the zeroth law of thermodynamics relative to the establishment of thermodynamic equilibrium between the two subsystems, say, A and B, with probability distributions $p_i(A)$ and $p_j(A)$, respectively[14]. In order to reach the necessary zeroth law, the following hypotheses was made. 1) If the probability distribution is complete and normalized with $\sum_{i=1} p_i = 1$ for the

IOP Conf. Series: Journal of Physics: Conf. Series **1113** (2018) 012013 doi:10.1088/1742-6596/1113/1/012013

composite system and the subsystems, the joint probability of the composite system should be given by $p_{ij}^{\alpha}(A+B) = p_i^{\alpha_a}(A)p_j^{\alpha_b}(B)$ where α, α_a and α_b are the incompleteness exponents of the composite system, subsystem A and subsystem B, respectively[17]. 2) If the probability distribution is incomplete and normalized with $\sum\limits_{i,j} p_{ij}^{\alpha}(A+B) = 1$, $\sum\limits_{i} p_i^{\alpha_a}(A) = 1$ and $\sum\limits_{j} p_j^{\alpha_b}(B) = 1$, the joint probability of the composite system should be given by $p_{ij}(A+B) = p_i(A)p_j(B)$[16].

In those works, the expressions of the joint probability are nevertheless hypothetical in order to get coherent results concerning the existence of the zeroth law in those thermostatistics formalisms. In what follows, we will show that the relationship of the incomplete joint probability mentioned above is a natural consequence of the composition of the fractals with independent scale factors.

Now imagine two multifractal phase spaces A and B in which we have the following incomplete normalizations :

$$\sum_i p_i^{\alpha_a} = 1 \tag{23}$$

and

$$\sum_j p_j^{\alpha_b} = 1. \tag{24}$$

where the sums are over all the segments of the two fractals, respectively.

Suppose now that A and B compose a composite multifractal C in the way described in section 3.4, we naturally have $r_{i,j} = r_i r_j$ and, at k stage, $\delta_{i_k,j_k} = \delta_{i_k}\delta_{j_k}/L_0$ or $p_{i,j} = p_i p_j$. From Eq.(19), we can write

$$\sum_{i,j} p_i^{\alpha} p_j^{\alpha} = 1$$

or

$$\sum_{i,j} p_{ij}^{\alpha} = 1$$

which is the incomplete normalization for the composite systems. Hence the hypothesis of the incomplete joint probability made in the context of nonextensive statistics[16] is supported from the point of view of fractal composition presented in this work.

If a system S is composed of m independent subsystems $\{A_1, A_2, \ldots, A_m\}$ with $\{l_1, l_2, \ldots, l_m\}$ the indices of the probability distributions of the subsystems, from Eq.(19), we can write

$$p_{l_1,l_2,\ldots,l_m}(S) = p_{l_1}(A_1)\,p_{l_2}(A_2)\ldots p_{l_m}(A_m)$$

and

$$\sum_{l_1,l_2,\ldots,l_m} (p_{l_1,l_2,\ldots,l_m})^{\alpha} = \sum_{l_1} p_{l_1}^{\alpha} \sum_{l_2} p_{l_2}^{\alpha} \cdots \sum_{l_m} p_{l_m}^{\alpha} = 1.$$

6. Conclusion

In this work, we have studied the composition of fractals. The composition is performed by alternating the iterations of the component fractals in a certain manner. We have studied composition with several possible alternations and calculated the dimension of the composite fractals.

IOP Conf. Series: Journal of Physics: Conf. Series **1113** (2018) 012013 doi:10.1088/1742-6596/1113/1/012013

- In the case of the composition of different fractals, the composite dimension can be expressed in the form of barycenter or average of component dimensions.
- For multifractals, the composite dimension can be expressed in a generalized Moran equation which can be solved numerically to find the composite dimension.
- It is proved that, in general, the composite dimension is bounded by the minimum and maximum of the component dimensions, and independent of the order of different substages in a stage.

We would like to mention that the composite dimensions found in this work are true for fractals which do not have overlapping segments or where there is no elements which intersect themselves during iteration. If there is overlapping, the composite dimensions presented here constitute upper bounds for the real dimension of overlapping figures. The difference between the real dimension and the upper bound depends on the degree of intersection of the shape.

In this work we have focused on the composition by simple and regular alternation of the iterations of the component fractals. The same method can be extended to more complicated cases where the number of substages changes with the increasing iteration or where the occurrence of substages is probabilist [5]. This work is in progress. We hope that the approach of composition (or decomposition) of fractals can find application in the study of fractal geometry as well as in the modeling of different structures and processes in nature.

References

[1] Mandelbrot Benoit B 1982 The Fractal Geometry of Nature (New York: W. H. Freeman and Co.)

[2] Mandelbrot Benoit B 1989 Les objets fractals (Flammarion)

[3] Falconer K 2003 Fractal Geometry: Mathematical Foundations and Applications 2nd ed. (Wiley)

[4] Le Méhauté, A, Nigmatullin R, Nivanen L, 1990 Fléches du temps et géométrie fractale (Hermes)

[5] Barnsley M, Hutchinson J E and Stenflo O, 2008 V-variable fractals and superfractals V-fractals: fractals with partial self-similarity, Advances in Mathematics **2189** 2051-2088

[6] Barnsley M, Hutchinson J E and Stenflo O 2012 V-variable fractals: dimension results, Forum Math. **24** 445-470

[7] Feder J 1988 Fractals (Plenum Press, New York)

[8] Wang Q A 2001 Incomplete statistics and nonextensive generalization of statistical mechanics, Chaos, Solitons and Fractals, **12** 1431-1437

[9] Wang Q A 2002 Nonextensive statistics and incomplete information, Euro. Phys. J. B **26** 357

[10] Wang Q A, 2004 Incomplete information and fractal phase space, Chaos, Solitons & Fractals **19** 639 cont-mat/0207647

[11] Wang Q A, Le Méhauté A, Nivanen L, Pezeril M, 2004 Fractal geometry, information growth and nonextensive thermodynamics, Physica A **340** 117

[12] Wang Q A, Le Méhauté A, 2004 Measuring information growth in fractal phase space, Chaos, Solitons & Fractals, **21** 893 cont-mat/0305540

[13] Wang Q A, Nivanen L, El Kaabouchi A, Badiali J P, Le Méhauté A, 2007 Incomplete normalization of probability on multifractals, Physica A **386** 92-100

[14] Wang Q A, Nivanen L, Le Méhauté A, Pezeril M, 2004 Temperature and pressure in nonextensive thermostatistics, 2004 Europhys. Lett., **65** 606

[15] Li W, Wang Q A, Nivanen L, and Le Méhauté A, 2005 On different q-systems in nonextensive thermostatistics, Euro. Phys. J. B, **48** 95

[16] Nivanen L, Pezeril M, Wang Q A, Le Méhauté A, 2005 Applying incomplete statistics to nonextensive systems with different q indices, Chaos, Solitons & Fractals **24** 1337

[17] Barnsley M and Vince A, 2011 The chaos game on a general iterated function system, Ergod. Th. & Dynam. Sys. **231** 1073-1079

IOP Conf. Series: Journal of Physics: Conf. Series **1113** (2018) 012014 doi:10.1088/1742-6596/1113/1/012014

An agent-based modeling for housing prices with bounded rationality

Weiping Wang[1,2,3], Saini Yang[1,2,3], Fuyu Hu[4], Zhangang Han[5] and Carlo Jaeger[2]

[1]Key Laboratory of Environmental Change and Natural Disaster, Ministry of Education, Beijing Normal University, Beijing, China

[2]State Key Laboratory of Earth Surface Processes and Resource Ecology, Beijing Normal University, Beijing, China

[3]Faculty of Geographical Science, Academy of Disaster Reduction and Emergency Management, Beijing Normal University, Beijing, China

[4]School of Civil Engineering and Environmental Science, University of Oklahoma, Norman,USA

[5]College of System Science, Beijing Normal University, Beijing, China

E-mail: `yangsaini@bnu.edu.cn`

Abstract. This study proposes an agent-based model to investigate major stakeholders' behaviors in the housing market. The proposed model mimics the heterogeneous behaviors of individual buyers and sellers in a housing market considering bounded rationality. The simulation results of case study in Shanghai are robust and reproduce stylized facts including as volatility clustering, absence of autocorrelations, heavy tail, loss asymmetry, and aggregational gaussianity on the absolute return.

1. Introduction

Research into the times series of financial returns demonstrate the existence of typical statistical properties of asset returns, such as distributional properties, tail properties and extreme fluctuations, leverage effects and downside correlations, pathwise regularity, and the linear and nonlinear dependence of returns [1, 2, 3, 4]. These statistics are also known as stylized facts, which deal with the difficulty of justifying the level of large variability (positive and negative) in asset returns in terms of fundamental economical variables.

In China, the real estate business plays a crucial role in the economy. Housing prices and its inflation illustrate the stylized facts of financial time series[5]. Simutaneously, the housing market features heterogeneity and bounded rationality.

Kain[6] pointed out that the usefulness of a housing market theory explicitly depends on how it deals with the heterogeneity that influences the behavior of multiple stakeholders in the housing market. Especially in China, researchers should pay more attention to the inter- and intra-municipality differences in housing because of the country's significant heterogeneity in its emerging urban housing market [7]. A better description of the behaviors in the housing market require a clear understanding of the nature and implications of this heterogeneity.

Malpezzi and Wachter (2005) claimed that in illiquid markets, such as in real estate, prices may be very volatile, displaying dramatic price variations that are extremely difficult to explain

IOP Conf. Series: Journal of Physics: Conf. Series **1113** (2018) 012014 doi:10.1088/1742-6596/1113/1/012014

using rational theory. Rational behavior theory implies that the market participants know the all other participants' beliefs, which seems excessively unrealistic [8]. Arthur (1994) pointed out that this is infeasible, stating "the type of rationality we assume in economics, perfect, logical, deductive rationality, is extremely useful in generating solutions to theoretical problems. But it demands much of human behavior, much more in fact than it can usually deliver" [9]. This provides a new perspective to modeling market behaviors: an inductive approach with bounded rationality, which refers to the phenomenon that individual rationality in decision-making is limited by the availability of information, the cognitive limitations of minds, and the limited time to make decisions [10]. The idea of bounded rationality was widely used in decisions behind policies[11], and economics and finance [12] since its introduction.

With the development of computational capacity and the availability of big data, financial markets and researchers have used computer-based models, such as the agent-based model. Ferreira proposed a grand canonical minority game agent-based model that reproduces stylized facts including volatility clustering, fat tails, uncorrelated returns, and slowing decay [13]. Similarly, Challet presented and studied a minority game-based model of a financial market with adaptive agents reflecting some stylized facts such as the fat tailed distribution of returns and volatility clustering [14]. Lux build a multi-agent model of a financial market with two groups of agents adopting differing strategies to value assets and making decisions (purchase, sell, hold) based on market prices and their own appraisals [15]. These studies reveal promising potential for intelligent computing models, such as an agent-based model, in financial markets ([16]; [17]; [18]).

Thus, this study aims to develop a new agent-based model to mimic the heterogeneous individual behaviors of major stakeholders (sellers and buyers) in a housing market that includes bounded rationality to capture the systematic behavior of the housing market and reproduce stylized facts. This paper is organized as follows. In the next section, we introduce the proposed model. In Section 3, we describe the simulation of our proposed model and discuss the results in section 4. Finally, we present conclusions in Section 5.

2. The model

Psychology researchers indicate that when humans face complicated problems, they tend more toward looking for patterns rather than using deductive logic, that is, "as feedback from the environment comes in, we may strengthen or weaken our beliefs in our current hypotheses, discarding some when they cease to perform, and replacing them as needed with new ones" [9]. This process coincides with the agent-based modeling process, in which each agent starts from simplified models, learns from the feedback, and adapts the model over time. The stakeholders in a real-world system usually have no access to complete information. Therefore, people make decisions based on the information they can obtain. When making decisions, people integrate limited historical information with their expectations for the future market with bounded rationality, heavily influenced by personal preferences and emotional sensitivity.

The proposed housing market model consists of N interacting agents, where Nb are buyers and the remaining $Ns=N\text{-}Nb$ are sellers. The agents act in a given region R (e.g. a rectangular region sized by $X*Y$) and interact with other agents in the neighborhood. We describe each detail of the proposed model in the following.

Agents

There are two types of agents: sellers and buyers. Sellers have two types of activities: selling and holding. In each time interval, the seller's decision about the housing price is based on the costs and expected profit. Here, the expected profit is not derived from theoretical models, but based on the individual's memory of limited historical information and bounded rationality. The seller will make a sale if the buyer offers a price meeting their proposed price. Otherwise, the seller will hold. Buyers either purchase or hold. Buyers have similar decision mechanisms,

IOP Conf. Series: Journal of Physics: Conf. Series **1113** (2018) 012014 doi:10.1088/1742-6596/1113/1/012014

and will decide upon the upper bound of the price based on memories of limited historical information and bounded rationality. Only when the seller proposes a price that is no higher than buyer's upper bound will the buyer make the purchase.

Neighborhood

Our model assumes incomplete information. A seller or buyer cannot obtain all information about the market, but rather relies on information from close associates (neighborhood). This closeness includes both physical distance or the social distance in a social network. Incomplete information also means that the historical information, such as prices and volumes, expires after a certain time. Therefore, it is important to define each agent's "neighborhood" that supplies the agent with information. The model also requires a definition for information expiry, meaing that the information is deleted from the memory pool and is invalid for future decision-making.

Pricing Mechanism

The pricing mechanism for a certain commodity varies with people, location, and time. It is stochastic even for one person at a given time and location. To avoid loss of generality, we set a pool of strategies for agents in the proposed model.

Option Pool

Bounded rationality means that agents will not theoretically deduce the so-called "optimal option" (though it may not be optimal as other agents dynamically evolve). Instead, agents will select an option from an "option pool". As an example, a option pool for a seller can be:

(i) The price with the most sales according to memory in the last 12 months;
(ii) The price most suitabile to the neighborhood in the last 12 months;
(iii) The weighted average price in the neighborhood in the last 12 months;
(iv) The maximum of the prices of options 1-3.

The suitability of a price S_i is the number of purchases at this price. The weighted average price in neighborhood P_{mean} can be calculated by Eq. (1).

$$P_{mean} = \frac{\sum_{i=1}^{Nb} S_i P_i}{\sum_{i=1}^{Nb} S_i} \tag{1}$$

where Nb indicates the total number of buyers and S_i represents the price offered by the ith buyer.

Similarly, buyers will decide their upper bound of offered prices based on limited historical information. Here, we propose 6 buyer strategies:

1. The same price as in the last month;
2. The average price in the last 12 months;
3. The projected price based on the last 12 months;
4. The average price in the last 4 months;
5. The average price in the last 6 months.

The probability an agent selects an option from among their option pool follows a probability distribution that will evolve over time.

Evolution mechanism

Perception will heavily influence decision making. The perception of a "good" price or "good" strategy will increase the possibility that this price or option will be used in the future. In the beginning, we may have no preference for a certain option, but the preference will develope and the probability of using a specific option will change based on evaluations over the previous time interval. Buyers and sellers will increase or decrease the probability of selecting an option based on the perception of the option, and judgments of how promising this option is.

For example, in the first time interval, an agent sets the selection probability of the 4 strategies at (1,1,1,1). After the first time interval, the agent finds that the second strategy is the one closest to the deal price. He would then change the selection probability in the second time interval to (1,2,1,1). Thus, the second strategy will have a better chance of selection in the second time interval.

Emotional factor

Humans are emotional. If a buyer has a demand and cannot not be satisfied for a long time, he will be influenced by the number of consecutive time intervals with unsatisfied demand, and is willing to increase the purchasing price. Assuming the emotional factor conforms to a sigmoid function, if the ith buyer has unsatisfied demand of N in the last month, the emotional factor is expressed as in Eq. (2).

$$E_i = \frac{K * P_i}{1 + e^{Nt}} \tag{2}$$

Here, K is a multiplier and t indicates the number of consecutive months with unsatisfied demand.

Compared to the standard economic theory assuming that agents are rational and often try to pursue maximum profit in the housing market [19], a significant difference in this model is that each agent will have incomplete information and make random selections based on bounded rationality.

3. SIMULATIONS

We developed a simulation platform with MATLAB 2012b to illustrate the simulation results of the case study under different scenarios.

We use the real-world monthly average housing prices for Shanghai in 2013 as the input for the initial condition: 27,369; 27,399; 25,826; 26,452; 26,552; 26,635; 26,692; 26,913; 26,927; 2,7053; 27,464; and 27,508 (unit: RMB/m^2)(Source: www.soufun.com)

In our simulation, there are N=100 agents, Nb=80 buyers and Ns=20 sellers, randomly located in a a (100, 100) lattice network, as illustrated in Figure 1. The Euclidean distance determines the neighborhood size. Two agents within Euclidian distance D are defined as neighbors. To test the influence of information incompleteness, we consider 5 values for D: 25, 50, 75, 100, 125, and 150. The threshold of D is based on the diagonal distance of the lattice network, which is 100*sqrt(2) in our case study. A seller must estimate buyers' future demand to make sure to meet the demand and reduce stocking costs. Thus, there will be a lag of Δt for sellers, meaning that houses will be ready for sale after construction time Δt (Here, we set $\Delta t = 1$). Each buyer has a demand, uniformly distributed across (1,10). The strategy pool and evolution mechanism for sellers and buyers are as discussed in Section 2.

The simulation is terminated after reaching iteration number tn=1000.

4. Results and analysis

4.1. Stylized facts

To test and analyze the simulated results of our model, we defined the price return *r(t)* as:

$$r(t) = \log P_t - \log P_{t-\delta} \tag{3}$$

Where, P_t denotes the price at time t; and δ is the time delay of no less than 1.

For financial time series, the major statistical properties of asset returns are described using distributional properties, tail properties and extreme fluctuations, pathwise regularity, and linear and nonlinear dependence of returns in time and across stocks [1]. Thus, we focus on these features to analyze our simulation results in terms of price returns *r(t)*.

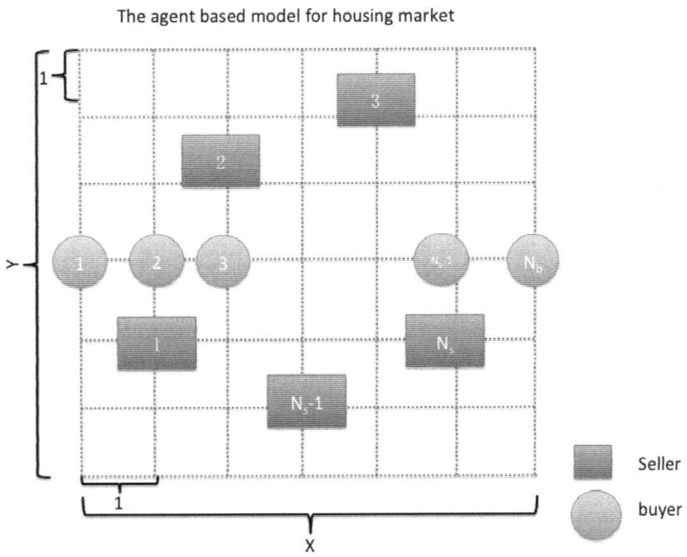

Figure 1. The agent based model of the housing market

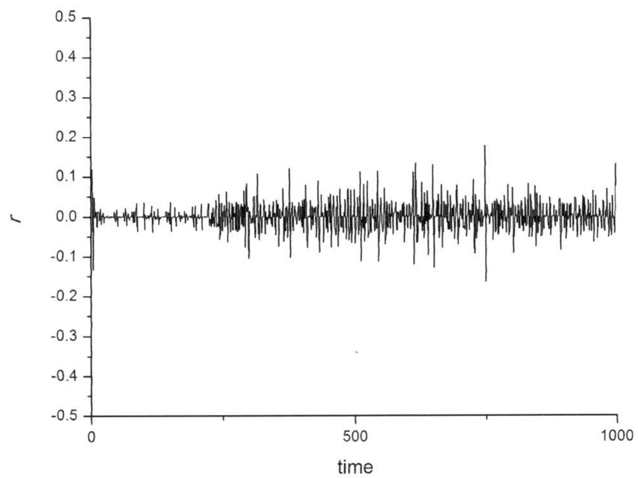

Figure 2. Time series of housing price returns

After 1,000 iterations, we obtain a time series of returns for simulated housing prices, as shown in Figure 2. This housing price time series illustrates 6 stylized facts of financial assets.

Absence of autocorrelations

The return autocorrelation $C(\delta)$ can be defined as:

$$C(\delta) = \frac{< r(t+\delta)r(t) > - < r(t+\delta) >< r(t) >}{r(t)^2} \quad (4)$$

Where $< ... >$ denotes the time average of price returns. Figure 3 shows that there is a slow

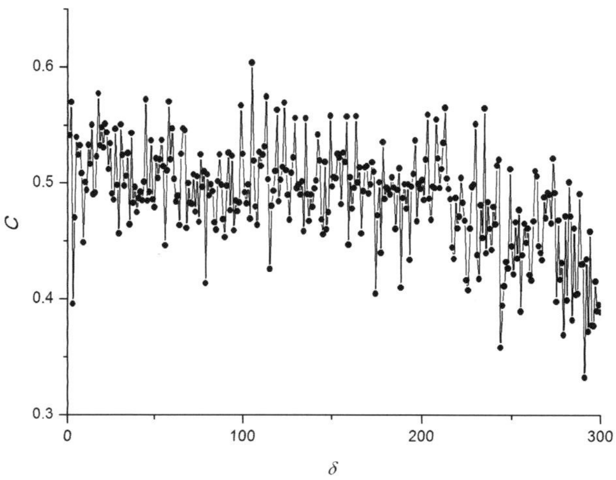

Figure 3. Autocorrelation (dotted line) as a function of δ

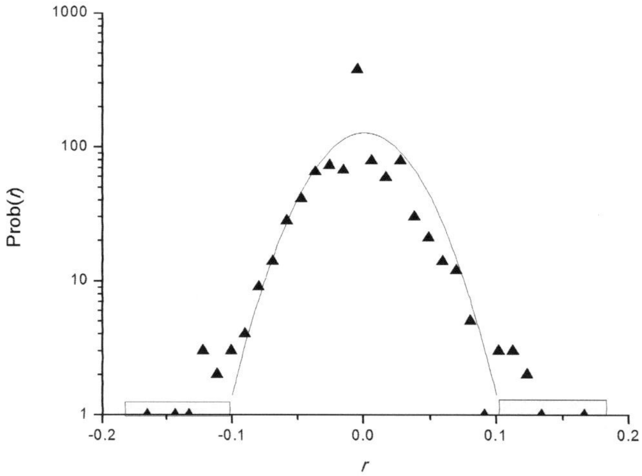

Figure 4. Comparison of probability distribution of returns and fitted normal distribution.

decay decay in C that can be interpreted as a long-range correlation in the returns. This is evidence of a sort of memory in the series.

Heavy tails

To illustrate the heavy tail of the returns calculated from the simulated results, we plot the logarithmic value of the probability mass function (PMF) of r in solid triangles and the logarithmic value of the PDF for the fitted normal distribution in a solid line in Figure 4. The figure shows that the probability distribution of the simulated results exhibits a fat tail, as indicated in the yellow zones.

Loss asymmetry

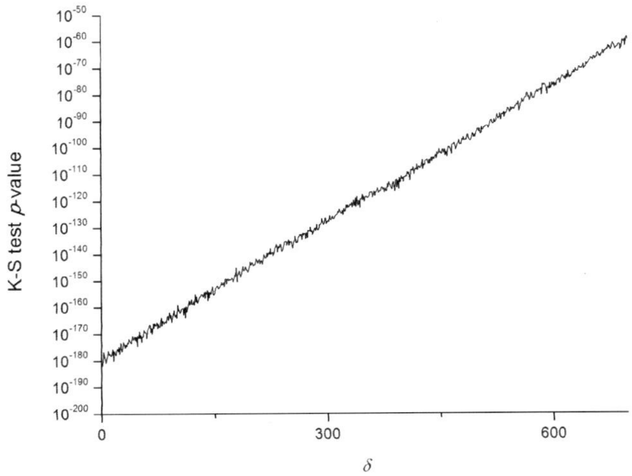

Figure 5. Kolmogorov–Smirnov test (K-S test) p–value as a function of δ

Loss asymmetry is the phenomenon wherein there are significantly large downward movements in the time series but not equally large upward movements [1].

Kurtosis measures how outlier-prone a distribution is. The kurtosis of a normal distribution is 3, and more more outlier-prone distributions have kurtosis greater than 3.

The kurtosis of our simulated result is 6.2803, indicating that the simulated results show loss asymmetry.

Aggregational Gaussianity

When increasing the time scale over which we calculate returns, the distribution of the price returns look increasingly like a normal distribution, indicating aggregational Gaussianity [1]. The Kolmogorov–Smirnov test determines whether a price return distribution fits a standard normal distribution. As shown in Figure 5, when the time delay increases, the p–value increases correspondingly, implying that the corresponding price return distribution is closer to a normal distribution.

Volatility clustering

Volatility clustering has a quantitative signature, that is, large price variations probably have subsequent large price variations.

In Figure 2, we observed this phenomenon in the time intervals [280,320] and [360 380].

These stylized facts prove that the simulation results from the proposed model possess common features of financial time series. The flexibility of this agent-based model can effectively represent the complexity and adaptive capacity of the housing market. Thus, we can apply this model to analyze other systems with similar characteristics or further explore the complex behaviors in those systems.

4.2. Incompleteness of information

Incomplete information is one root cause of bounded rationality. To explore the influence of information incompleteness on our model, we performed a sensitivity analysis of the neighborhood size D using a range of 25 to 150 with a step size of 25.

When the Euclidian distance is set to be 25, the logarithmic value of the PMF of the price return shows heavy tails with a Kurtosis of 40.5082, volatility clustering, and loss asymmetry

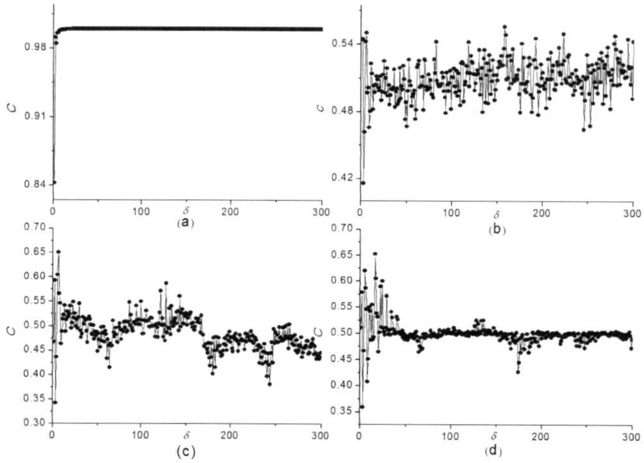

Figure 6. Euclidian distance at 25 (a), 75 (b), 125 (c), and 150 (d) where the absence of autocorrelation does not hold.

with skewness of -3.6175. Autocorrelation (dotted line) as a function of δ shows that C gradually approaches and stabilizes around 1, indicating the obvious presence of autocorrelation. The Kolmogorov- Smirnov test for price returns shows that p equals 0. However, both the absence autocorrelation and the aggregational Gaussianty do not hold when the Euclidian Distance is 25.

When we increase the Euclidian distance to 50, volatility clustering also exists, with Kurtosis equal to 7.3095, showing heavy-tails. Skewness equals 0.3338, indicating loss asymmetry. However, for autocorrelation, we observe a slow increase in C, which does not meet the stylized facts of financial assets.

When the Euclidian distance is set at 75, 125, and 150, respectively, the stylized facts of financial assets on volatility clustering, heavy tails, and loss asymmetry hold, with Kurtosis equal to 7.1207, 22.2322, and 63.7230; and skewness equal to -0.1527, -0.0090, and 0.4480, respectively. For autocorrelation, we observe a fluctuation around 0.5 with no trend of decay or increase when the Euclidian distance is 75. The autocorrelation first decays and then increases and declines again when the Euclidian distance is set at 125,. When Euclidian distance is set at 150, the autocorrelation first decreases and then increases to finally reach a smooth tread near 1, as shown in in Figure 6.

Table 1 shows stylized facts with different neighborhood sizes, where Y indicates that the stylized fact holds with given D, and N indicates that it does not hold. The results show that aggregational Gaussianity and the absence of autocorrelation are more sensitive to the neighborhood size, which represents the incompleteness of information.

5. Conclusion

This study aimed to create an agent-based model to mimic the heterogeneous individual behaviors of major stakeholders (seller and buyers) in a housing market that considers bounded rationality. The model can capture the systematic behavior of the housing market, and the

Table 1. Stylized facts with different neighborhood sizes D.

D	Absence of autocorrelation	Heavy tails	Loss asymmetry	Aggregational Gaussianity	Volatility clustering
25	N	Y	Y	N	Y
50	N	Y	Y	N	Y
75	N	Y	Y	N	Y
100	Y	Y	Y	Y	Y
125	N	Y	Y	N	Y

statistical properties of the data generated by a simulation are in line with the stylized facts expected from the financial data, such as volatility clustering, heavy tails, and loss asymmetry. etcetera. It is especially noteworthy that the aggregational Gaussianity and the absence of autocorrelation are sensitive to the incompleteness of information. This model provides an applicable tool to analyze complicated systems with incomplete information and heterogenous behaviors.

References

[1] Cont R 2001 *Quantitative Finance* **1** 223–236
[2] Bouchaud J P and Potters M 2001 *Physica A* **299** 6070
[3] Xavier Gabaix P G and Vasiliki Plerou H E S 2003 *Nature* **423** 267–270
[4] Mantegna R N and Stanley H E 1995 *Nature* **376** 46–49
[5] Zhang C 2013 *Journal of Policy Modeling* **35** 75–87
[6] Kain J F and Quigley J M 1975 *Housing Markets and Racial Discrimination: A Microeconomic Analysis* (National Bureau of Economic Research, Inc)
[7] Zhou Y 2006 *Habitat International* **30** 277304
[8] Esmaeili M, Vancheri A and Giordano P 2013 *Emergent Phenomena in Housing Markets*
[9] Arthur W B 1994 *American Economic Review* **84** 406–11
[10] Simon H A 1991 *Organization science* **2** 125–134
[11] Stadelmann D and Torgler B 2013 *Plos One* **8** 1–8
[12] Shubik M 1997 *Journal of Economics* 1646
[13] Ferreira F F, Oliveira V M D, Crepaldi A F and Campos P R A 2005 *Physica A* **357** 534542
[14] Challet D, Marsili M and Zhang Y C 2001 *Physica A* **294** 514 – 524 ISSN 0378-4371
[15] Lux T and Marchesi M 1998 *Nature International Weekly Journal of Science* **397** 498–500
[16] Ettema D 2011 *COMPUTERS ENVIRONMENT AND URBAN SYSTEMS* **35** 1–11
[17] Magliocca N, Safirova E, McConnell V and Walls M 2011 *COMPUTERS ENVIRONMENT AND URBAN SYSTEMS* **35** 183–191
[18] Lye R, Tan J P L and Cheong S A 2012 *PHYSICA A-STATISTICAL MECHANICS AND ITS APPLICATIONS* **391** 5521–5531
[19] Paraschiv C and Chenavaz R 2011 *HOUSING STUDIES* **26** 329–352

Acknowledgement

This study was supported by the National Key Program of China (No.2016YFA0602403), the fund for Creative Research Groups of the National Natural Science Foundation of China (No.41321001), scholarship from China Scholarship Council (CSC.201606040133), Chengdu Science and Technology Bureau(2015-HM01-00564-SF), and the ICCR-DRR at BNU. We also thank the reviewers for their valuable suggestions to improve the paper.

Comparative study of discrete Boltzmann model and Navier-Stokes

Yudong Zhang[1,2], Aiguo Xu[1,3,*] and Guangcai Zhang[1]

1 Laboratory of Computational Physics, Institute of Applied Physics and Computational Mathematics, Beijing, 100088, China

2 Key Laboratory of Transient Physics, Nanjing University of Science and Technology, Nanjing 210094, China

3 Center for Applied Physics and Technology, MOE Key Center for High Energy Density Physics Simulations, College of Engineering, Peking University, Beijing 100871, China

E-mail: Xu_Aiguo@iapcm.ac.cn

Abstract. Discrete Boltzmann model (DBM) is a type of coarse-grained mesoscale kinetic model derived from the Boltzmann equation. Physically, it is roughly equivalent to a hydrodynamic model supplemented by a coarse-grained model for the relevant thermodynamic non-equilibrium (TNE) behaviours. The Navier-Stokes (NS) model is a traditional macroscopic hydrodynamic model based on continuity hypothesis and conservation laws. In this study, the two models are compared from two aspects, physical capability and computational cost, by simulating two kinds of flow problems including the thermal Couette flow and a Mach 3 step problem. In the cases where the TNE effects are weak, both the two models give accurate results for the hydrodynamic behaviour. Besides, DBM can provide more detailed non-equilibrium information, while the NS is more efficient if concern only the density, momentum, energy and their derived quantities. It is concluded that, if the TNE effects are strong or are to be investigated, the NS is insufficient while DBM is a good choice. While in the cases where the TNE effects are weak and only the macro flow fields are to be studied, the NS is more preferable.

1. Introduction

In recent years, discrete Boltzmann model (DBM) has been well developed to investigate various of complex flow situations where the thermodynamic non-equilibrium effect (TNE) is significant [1–3]. It provides a new point of view to study the non-equilibrium characteristics of flow system. Up to now, DBM has been successfully used in non-equilibrium phase transition and multiphase flow [4], flow instability [5–7], reactive flow [8, 9], shock and detonation [10–12], etc.

Compared with traditional hydrodynamic model, the Navier-Stokes (NS) equation [13, 14], DBM has more kinetic information. Generally, there are at least three different aspects between the two models. Firstly, the evolution equations of DBM are a set of the discrete particle velocity distribution, while the evolution equations of NS are those for the conservation laws of mass, momentum, and energy of control volume. Secondly, apart from the discretization of time and space, the discretization of particle velocity space is necessary for DBM. As a result, the actual number of evolution equations of DBM is determined by the discrete velocity model and usually is more than that of the NS. Thirdly, the numerical stability of DBM needs also consider the discrete velocities besides the general Courant–Friedrichs–Lewy (CFL) condition [14]. Nevertheless, DBM is necessary in the following two cases: (i) the TNE is so strong that the NS does not work well any more, or (ii) not only the hydrodynamic behavior but also the relevant TNE effect are to be investigated.

Since each model has its own advantages and limitations. In this work, a comparative study of DBM and NS is made. Two kinds of problems, including thermal Couette flow and Mach 3 step problem, are simulated. The two models are both used under the same conditions. The physical capabilities and computation times are compared.

2. Model and equations

2.1. Discrete Boltzmann model

The evolution equation of DBM read

$$\frac{\partial f_i}{\partial t} + v_{i\alpha} \cdot \frac{\partial f_i}{\partial r_\alpha} = -\frac{1}{\tau}(f_i - f_i{}^{+}), \tag{1}$$

where f_i^+ can be chosen as Maxwell distribution, Ellipsoidal Statistical (ES) distribution, and Shakhov distribution which correspond to BGK, ES-BGK, and S-model, respectively. According to the modeling of DBM, the discrete distribution function can be solved from the kinetic moments [15],

$$f_i^+ = C_{ij}^{-1}\hat{f}_j, \tag{2}$$

where C_{ij}^{-1} is the coefficient matrix of discrete velocity and $\hat{f}_j$ is kinetic moment calculated from integral of the original distribution function.

The NTE quantities are defined as [1]

$$\Delta_m^* = \sum_i (f_i - f_i^{eq}) \underbrace{(v_{i\alpha_1} - u_{i\alpha_1}) \cdots (v_{i\alpha_m} - u_{i\alpha_m})}_{m}, \tag{3}$$

where f_i^{eq} is discrete Maxwell distribution.

2.2. Navier-Stokes equations

Navier-Stokes equations include continuity equation, momentum conservation equation, and energy conservation which read [13]

$$\frac{\partial \rho}{\partial t} + \frac{\partial(\rho u_\alpha)}{\partial r_\alpha} = 0, \tag{4}$$

$$\frac{\partial(\rho u_\alpha)}{\partial t} + \frac{\partial}{\partial r_\beta}\left[\rho(T\delta_{\alpha\beta} + u_\alpha u_\beta) - 2\mu\frac{\partial u_{<\alpha}}{\partial r_{\beta>}}\right] = 0, \tag{5}$$

$$\frac{\partial\left[\rho(e + \frac{u^2}{2})\right]}{\partial t} + \frac{\partial}{\partial r_\alpha}\left[\rho u_\alpha\left(e + T + \frac{u^2}{2}\right) - \left(\kappa'\frac{\partial T}{\partial r_\alpha} + 2u_\alpha\mu\frac{\partial u_{<\alpha}}{\partial r_{\beta>}}\right)\right] = 0, \tag{6}$$

where $\mu = \tau\rho T$ and $\kappa' = c_p\tau\rho T$ are coefficients of viscosity and heat flux, respectively. In fact, Navier-Stokes equation can also be derived from Eq.(1) by Champan Enskog expansion [16] under the first order approximation. The viscosity and heat flux are correspond to non-organizational momentum flux(NOMF) and non-organizational energy flux(NOEF) [12], respectively.

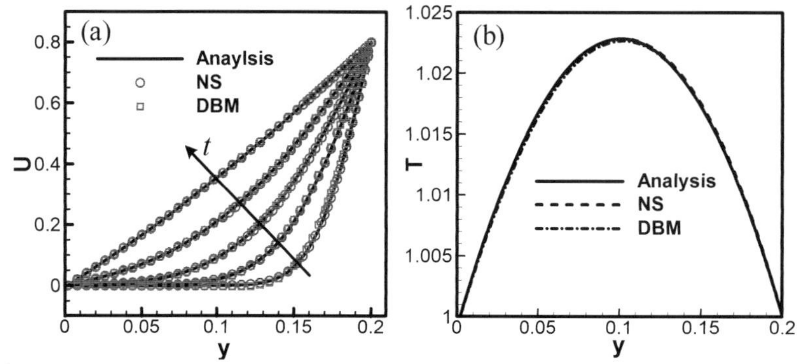

Figure 1. (a) The distribution of velocity over time; (b) The distribution of temperature (steady-state).

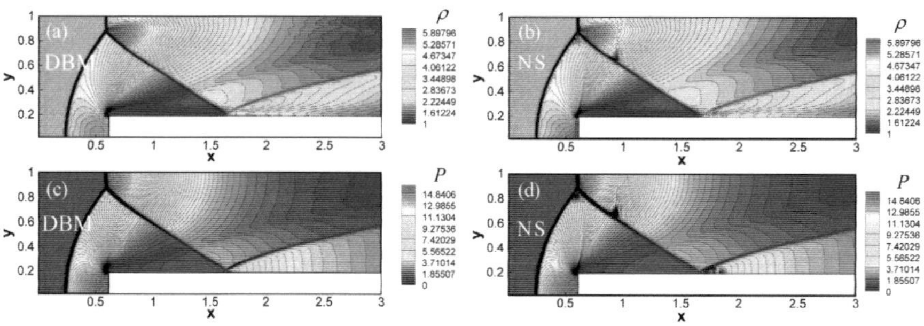

Figure 2. Contour maps of density and pressure: (a) density by DBM, (b) density by NS, (c) pressure by DBM, (d) pressure by NS.

3. Results and Analysis

3.1. Thermal Couette flow

Firstly, the thermal Couette flow is simulated by using DBM and numerical solving NS equations. The same time and space steps are adopted for two models. The space derivative is solved by second order upwind scheme in DBM and second center scheme in NS. The results are shown in Fig.1. Figure 1(a) shows the distribution of velocity at various time. Figure 1(b) gives the distribution of temperature when the flow reaches to its steady state. The results of two models are almost identical and both fit well with analysis solution.

3.2. Mach 3 step problem

As a classical two dimensional compressible flow problem, the Mach 3 step problem is widely used to verify various new models [17]. Here, we use DBM and NS simulate this problem under same condition. The NND scheme [18] is adopted to solve space derivative in both model. Figure 2 shows the contour maps of density and pressure by two models. The results are well consistent with each other. In addition, the NTE quantities provided by DBM are also shown in Fig. 3, which shows the non-equilibrium characteristics of the whole flow field. Four kinds NTE are displayed and each describes the non-equilibrium strength from different perspective. Those NTE quantities cannot obtained by NS.

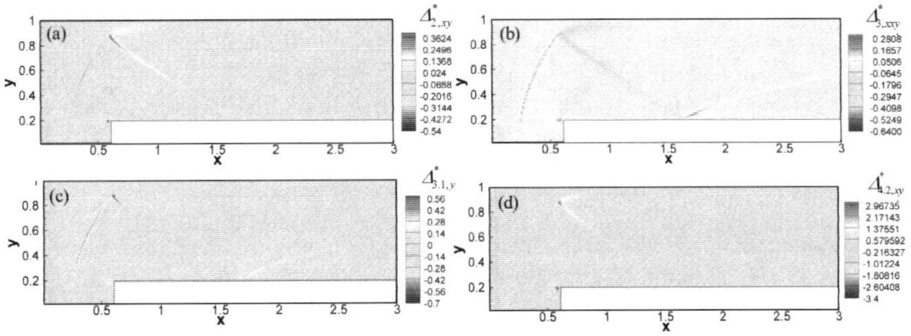

Figure 3. Contour maps of TNE quantities provided by DBM. (a)$\Delta^*_{2,xy}$, (b)$\Delta^*_{3,xxy}$, (c)$\Delta^*_{3.1,y}$, (d)$\Delta^*_{4.2,xy}$.

3.3. Computational efficiency comparison

From the previous two example, we have learned that both DBM and NS can well describe thermal flows and high speed compressible flows. Now, the computational efficiencies of the two models are compared. The computational mesh for thermal Couette flow and Mach 3 step problem are 4×100 and 300×100 , respectively. The computation times are shown in Table 1.The calculation is performed on a PC with Inter(R) Core(TM) i7-4790 CPU @3.60GHz

Obviously, NS has a higher efficiency than DBM. So, if the NS can meet the requirements, it is better to choose NS. However, it should be note that DBM is easier to program than NS.

Table 1. Computational time comparison between DBM and NS.

	Thermal Couette flow	Mach 3 step
DBM (s/1000 steps)	17.08	506.1
NS (s/1000 steps)	3.14	227.8

4. Conclusion

The comparative study of DBM and NS is made by two kinds of flow problems. In the cases where both the two models are suitable, the simulation results of the two models are well inconsistent with each other. But when it comes to computational efficiency, NS has an advantage over DBM if only the evolutions of generally hydrodynamic quantities (density, flow velocity, temperature, pressure and their derived variables) are to be investigated. Physically, the DBM can provide more detailed non-equilibrium information than the NS. In the cases where the TNE effects are strong or the TNE effects are to be investigated, the NS fails to work and the DBM is physically necessary. In the cases where the TNE effects are weak and one needs only the macro flow fields, the NS is more preferred for saving computational time, especially when simulating high dimensional problems.

Acknowledgments

Acknowledgement The work is supported by National Natural Science Foundation of China [under grant nos. 11475028 and 11772064] and Science Challenge Project (under Grant No. JCKY2016212A501 and TZ2016002).

References

[1] Xu A, Zhang G and Ying Y 2015 *Acta Physica Sinica* **64** 184701
[2] Xu A, Zhang G and Gan Y 2016 *Mechanics in Engineering* **38** 361–374
[3] Xu A, Zhang G and Zhang Y 2018 Discrete boltzmann modeling of compressible flows *Kinetic Theory* ed Kyzas G Z and Mitropoulos A C (Rijeka: InTech) chap 02 URL `http://dx.doi.org/10.5772/intechopen.70748`
[4] Gan Y, Xu A, Zhang G and Succi S 2015 *Soft Matter* **11** 5336
[5] Lai H, Xu A, Zhang G, Gan Y, Ying Y and Succi S 2016 *Phys. Rev. E* **94** 023106
[6] Chen F, Xu A and Guangcai Z 2016 *Front. Phys.* **11** 183–196
[7] Lin C, Xu A, Zhang G, Luo K H and Li Y 2017 *Physical Review E* **96** 053305
[8] Lin C, Luo K, Fei L and Succi S 2017 *Scientific Reports* **7** 14580
[9] Lin C and Luo K 2018 *Computers & Fluids* **166** 176 – 183
[10] Xu A, Lin C, Zhang G and Li Y 2015 *Physical Review E* **91** 043306
[11] Lin C, Xu A, Zhang G and Li Y 2016 *Combust. Flame* **164** 137–151
[12] Zhang Y, Xu A, Zhang G, Zhu C and Lin C 2016 *Combust. Flame* **173** 483–492
[13] Chen M 2002 *Fundamentals of viscous fluid dynamic* (Beijing: Higher education press)
[14] Fu D 2006 *Computational aerodynamics* (Beijing: Aerospace press)
[15] Gan Y, Xu A, Zhang G and Yang Y 2013 *EPL* **103** 330–337
[16] Chapman S and Cowling T 1970 *The mathematical theory of non-uniform gases: an account of the kinetic theory of viscosity, thermal conduction and diffusion in gases* (Cambridge: Cambridge university press)
[17] Woodward P and Colella P 1984 *J. Comput. Phys.* **54** 115–173
[18] Zhang H 1988 *Acta Aerodynamica Sinica* **6** 143–165

IOP Conf. Series: Journal of Physics: Conf. Series **1113** (2018) 012016 doi:10.1088/1742-6596/1113/1/012016

Scientific collaborations within a university: From the viewpoint of complex networks

Weiquan Dong, Yizhen Yu and Xingang Wang

School of Physics and Information Technology, Shaanxi Normal University, Xi'an 710062, China

E-mail: wangxg@snnu.edu.cn

Abstract. While the topological features of scientific collaboration networks have been well explored in literature, the mechanisms by which the researchers are assembled remain not very clear. Here, by a 6-year publication data of the Shaanxi Normal University (SNNU), we construct the collaboration network and, combining with the *additional information* of the research staffs obtained through survey and interviews (including department affiliation, research direction, seniority, acquaintanceship, etc), investigate the key factors that promote scientific collaborations. It is found that, although of a relatively small size, the constructed network possesses still the general features of large-scale complex networks, including the community structures, the heterogenous degree and weight distributions, the preferential attachment of the new links, etc. In exploring the assembly mechanisms of the collaboration patterns, it is further revealed that among the many factors influencing scientific collaborations, the closeness of the research directions has a clear advantage over the others. The findings provide insights into the formation of the collaboration patterns within universities, and might helpful to the network modeling and research management in practice.

1. Introduction

Bibliometric studies over the past two decades have shown a rapid increase in the number of co-authored papers in every scientific discipline, as well as within and across different countries and geographic areas. Indeed, with the increased complexity of the scientific problems (e.g., the brain research) and experimental facilities (e.g., the Large Hadron Collider), the demanding for large-scale, interdisciplinary scientific collaborations is more urgent than ever [1, 2, 3]. To understand the patterns of scientific collaboration and their impacts on knowledge production and discipline evolution, scientific collaboration itself has became an active research topic in the past decades [3, 4, 5, 6, 7, 8, 9, 10, 11, 12, 13, 14, 15, 16, 17, 18], and has attracted considerable attentions in a variety of disciplines such as information science, management science, psychology, sociology, philosophy, to name just a few. Recently, stimulated by the significant progresses achieved in network science [19, 20, 21], a new powerful method, namely the collaboration (co-authorship) complex network, has been proposed and employed in analyzing the scientific collaborations [4, 5]. In the general collaboration networks, authors are represented by nodes, and two authors are connected by a link if they have co-authored a paper. Thanks to the well-recorded electronic bibliographic datasets, large-scale collaboration networks have been reliably constructed and systematically investigated for a variety of disciplines in recent years, in which many novel properties of scientific collaborations have been disclosed, e.g., the small-world and scale-free (rich-club) features [22], the community (cluster) structures [5], the associativity property [23], the robust-yet-fragile feature [24], the abnormal weight distribution [13], the formation and evolution of research teams [14], etc.

Although collaboration networks provide a powerful tool in unveiling the patterns of scientific collaborations, the efficiency of this method has been largely limited by the simplified node (author) and link (collaboration) properties. To be specific, many intrinsic properties of the network nodes (e.g., the gender and age of the authors) and links (e.g., the acquaintanceship between the authors) have been overlooked in literature [25]. This limitation is rooted in the format of the available bibliometric datasets, which record only the information of the author name and affiliation. As scientific collaborations among researchers are largely affected by their personal attributes [14, 15, 26], to have a deeper understanding on the assembly mechanism of collaboration patterns, it is therefore necessary to take into account the author attributes as much as possible. Motivated by this, in the present work we incorporate the method of collaboration network with the conventional methods (survey and interview) [27], and investigate the collaboration patterns within a university. Specifically, by a 6-year bibliometric dataset of SNNU, we construct the scientific collaboration network, and, based on a survey on the personal attributes for all the recorded researchers (including discipline, research interest/direction, working experience, date of enrollment, affiliation, acquaintanceship, etc), explore the key factors that influence scientific collaborations inside SNNU. An interesting finding is that, among the investigated factors, the research direction has a dominant role in influencing the collaborations than others. This finding reflects to a certain degree the current situation of scientific collaboration inside the China universities, which might be helpful for the research management and policy design in universities of the similar level.

2. Topological properties

The bibliometric dataset used in the present work is based on the journal publications recorded by the Institute for Scientific Information (ISI) during the period from January 2009 to December 2014. To focus on only the high-quality collaborations among research staffs working on natural science within SNNU, we filter the bibliometric dataset by the following requirements: 1) SNNU is appeared as the first address; 2) The paper is published in journals of natural science; 3) The journals are listed in the 1st and 2nd divisions of the Journal Citation Reports [28]; and 4) The paper is co-authored by at least two staffs from SNNU (i.e., excluding the solo publications and papers co-authored by the supervisor and his/her students). After such a filtering, there are totally 1645 papers left in the dataset, which are collaborated by $N = 503$ staffs distributed in $M = 16$ different departments in the university. (During the investigated period, the journal list is slightly modified year by year. In obtaining the dataset, we include a journal into the list once it appears in the 1st or 2nd divisions during the period.)

Having obtained the bibliometric dataset, we next construct the scientific collaboration network following the conventional method of network construction [4, 5]. In constructing the collaboration network, we represent each staff by a node, and two staffs are connected by a link if they have co-authored a paper. We have in total $L = 935$ links in the network. The weight of the link, w_{ij}, is defined as the number of co-authored papers between staffs i and j. The nodes are colored according to the departments, i.e., staffs affiliated with the same department are of the same color. The topological structure of the collaboration network is presented in Fig. 1. Unlike the general complex networks where nodes are reachable to each other (e.g., the Internet) [19, 20], here it is seen that the collaboration network is fragmental and scattering – a feature commonly observed in scientific collaboration networks [11, 25]. In particular, the network is constituted by a giant cluster and many small motifs. The giant cluster contains 401 nodes and 842 links, with staffs coming from 12 different departments. The disconnected motifs are of various sizes, ranging from 2 to 11.

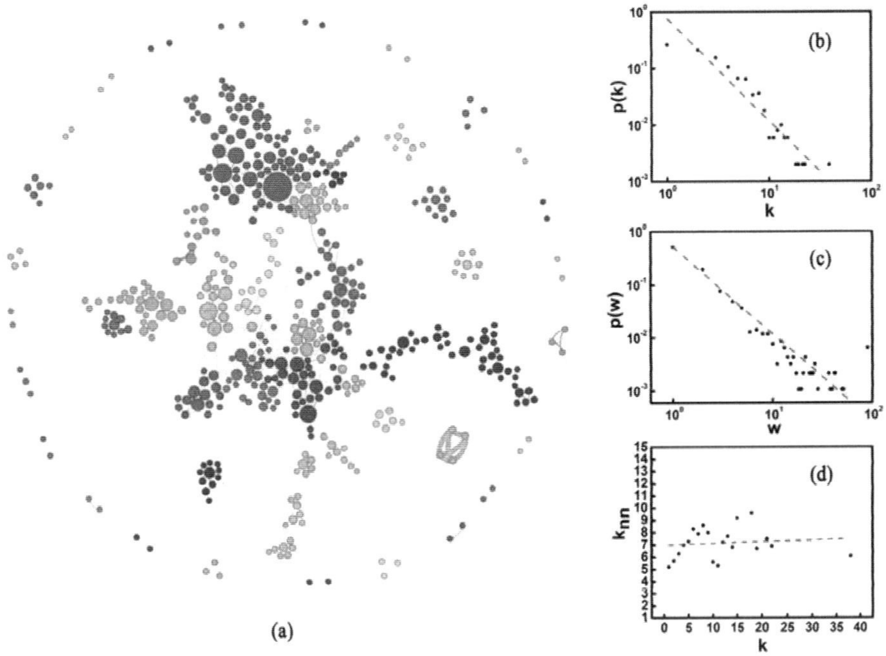

Figure 1. (Color online) The structure and topological properties of the scientific collaboration network. (a) The network structure, which consists of $N = 503$ nodes and $L = 935$ links. The node size is plotted as proportional to the node degree, and the edge thickness is plotted as proportional to the link weight (the number of co-authored papers). Nodes are colored according to the departmental information of the staffs. The network is made up of a giant cluster (which contains 401 nodes and 842 links), 12 small motifs (of size between 3 and 11), and 17 node-pairs. (b) The degree distribution follows roughly the power-law scaling, $p(k) \propto k^{\gamma}$, with the fitted exponent $\gamma \approx -1.8$. (c) The weight distribution follows roughly the power-law scaling, $p(w) \propto w^{\alpha}$, with the fitted exponent $\alpha \approx -1.6$. (d) The variation of the average-nearest-neighbor-degree, k_{nn}, as a function of k, with the fitted slope $\beta \approx 1.4 \times 10^{-2}$.

We go on to explore the topological properties of the network. Previous studies have shown that realistic complex networks normally possess the small-world and scale-free features [22]. As the network is fragmental, in investigating the small-world feature we focus on only the giant cluster. The numerical results show that the giant cluster has the averaged diameter $d \approx 6.4$ and the clustering coefficient $c \approx 0.55$. Clearly, the giant cluster possesses the small-world feature. To investigate the scale-free feature of the network, we plot in Fig. 1(b) the distribution of the node degrees, k. It is seen that the degrees are of heterogenous distribution. Particularly, the degree distribution follows roughly a power-law scaling within the range $k \in [1, 38]$, i.e., $p(k) \propto k^{\gamma}$, with the fitted exponent $\gamma \approx -1.8$. Besides the node degrees, the distribution of the link weights is also heterogenous, as depicted in Fig. 1(c). Specifically, the weight distribution follows roughly the power-law scaling $p(w) \propto w^{\alpha}$ within the range $w \in [1, 90]$, with the fitted exponent $\alpha \approx -1.6$. For scientific collaboration networks, another commonly interested topological property is assortativity (assortative mixing), i.e., the propensity that nodes of the similar degrees are connected to each other [23]. To explore this property, we measure the average-nearest-neighbor-degree, k_{nn}, for node of degree k in the network, and plot the variation of k_{nn} as a function of k [29]. If k_{nn} is increasing (decreasing) with k, then the network is classified as assortative (disassortative) [23]. The results are presented in Fig. 1(d). It is seen that, with the increase of k, the value of k_{nn} is fluctuating wildly, showing no clear trend of increase or decrease. (By a linear fitting of the data, in fact we have $k_{nn} \propto \beta k$, with the slope $\beta \approx 1.4 \times 10^{-2}$.) Fig. 1(d) thus indicates that the collaboration network of SNNU has nearly the neutral assortativity, which is different from the general collaboration networks studied in literature (which are normally assortative) [23, 29]. The neutral assortativity might be attributed to the limited size (resulting in the large fluctuations) and sparse connections (resulting in the fragmental structure) of the SNNU collaboration network.

In investigating scientific collaboration networks, a particularly interested topological property is the community structure [30], i.e., the network is partitioned into some communities, with the connectivity inside each community is much denser than between the communities. As community reflects the meso-scale structure of complex networks, it has been employed as an important approach to exploring the patterns of scientific collaborations in previous studies [4, 11, 15]. Here we choose the giant connective cluster in the network, and partition it into communities according to the algorithm proposed by Girvan and Newman [30]. In particular, we remove recursively the network links by a descending order of their betweenness, and monitor the variation of the network modularity, Q. [Roughly speaking, modularity measures the probability for a pair of connected nodes randomly chosen to be staying within the same community, which is varying between 0 (no community structure) and 1 (distinct community structures). Please refer Ref. [30] for the detail definition of Q]. By finding the maximum modularity, Q_{max}, we are able to identify the optimal partition of the communities. Numerically, we find that $Q_{max} \approx 0.9$, under which the network is partitioned into $M' = 21$ communities, with the size of the communities ranging from 3 to 62.

Do the identified communities reflect the realistic situation? In partitioning the staffs within a university into communities [30], a conventional approach in social science is to group the staffs according to their departmental affiliations. That is, staffs from the same department form a community. Apparently, the two types of communities, i.e., the collaboration and department communities, are different from each other. For instance, the giant cluster contains 21 communities, while the staffs are from only 12 departments. However, as scientific collaborations between researchers are largely influenced by their disciplines and acquaintanceship, the two types of communities do have certain degree of correlations. To quantify the correlation between the two types of communities, we introduce the community-correlation coefficient (CDC)

$$r = \langle r_l \rangle = \left\langle \frac{n'_l}{n_l} \right\rangle \tag{1}$$

with n_l the size of the lth collaboration community identified through the Girvan-Newman algorithm, and n'_l the maximum number of nodes affiliated to the same department within this community. $\langle \ldots \rangle$ represents that the result is averaged over all M' communities. Clearly, the larger r is, the stronger the correlation between the two types of communities is, and the stronger the community structures influenced by the departmental affiliations will be. The calculated result shows that $r \approx 0.8$. This strong correlation is somewhat expectable, as scientific collaborations within the same discipline in general have the higher propensity than cross disciplines, and, more importantly, staffs within the same department are usually acquaintances and friends [1, 2]. We would like to note that, while staffs within the same department are more likely to collaborate with each other (as compared with the collaborations between staffs from different departments), the collaborations are not of the equal probability, i.e., there exist some collaboration patterns within the department. For instance, staffs in the physics department are organized into 4 communities. Our above analysis of community structures thus suggests that departmental affiliation is influential, but not crucial to the scientific collaborations.

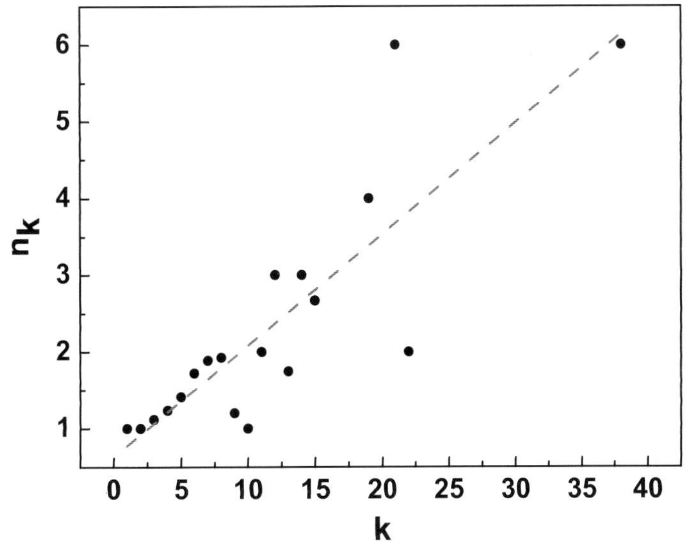

Figure 2. The preferential attachment mechanism of the newcomers in collaborating with the veterans. k: the degree of the veterans joined SNNU before 2009. n_k: the averaged number of newcomer-neighbors joined SNNU between 2009 and 2014. The fitted results show that $n_k \propto \eta k$, with $\eta \approx 0.15$.

3. Assembly mechanisms

To explore the key factors promoting scientific collaborations, we next study the collaboration patterns at the nodal level, based on the additional information of the staffs obtained through the conventional approaches of survey and interview. Specifically, we are going to analyze in detail: 1) the collaboration propensity of the newcomers, i.e., with whom the newcomers tend to collaborate; 2) the collaboration patterns within three specific, small-size departments; and 3) the collaboration pattern at the departmental level within SNNU.

In modeling the growth of collaboration networks, it has been commonly assumed that the new nodes (newcomers) are connected to the existing nodes (veterans) by the mechanism of preferential attachment [6, 9, 10, 16]. That is, the larger the degree of an existing node is, the higher the possibility for it to be connected by the newcomer will be. While preferential link attachment has been observed in the empirical data of many natural and man-made complex networks [19], it remains unknown whether it applies to the growth of scientific collaboration network, as the relevant data is unavailable from the electronic bibliographic datasets. This suspicion is reasonable, as it usually takes a certain period for a newcomer to familiarize the new academic environment and find the new collaborators, resulting in probably the non-preferential link attachment. According to the surveyed information, we identify from the dataset the veterans who joined SNNU before 2009, and plot in Fig. 2 the average number of newcomers (staffs joined SNNU between 2009 and 2014) they have collaborated with, n_k, versus their degrees, k. It is seen that, in accordance with the assumptions made in previous modeling studies, the value of n_k is indeed increasing with k by roughly a linear fashion. Fig. 2 provides (probably the first) empirical evidence showing the applicability of preferential attachment mechanism in modeling scientific collaboration networks.

We continue to analyze the collaboration patterns within a specific department. Due to the limited resources, we have only surveyed three research-oriented departments inside SNNU, namely the departments "A", "B", and "C". Department "A" is a national engineering laboratory, which was founded in 2008 and now has 41 incumbent staffs. According to our dataset, department 'A" has totally published 96 co-authored papers, which involve $N_A = 35$ staffs and give $L_A = 54$ collaboration links. The topological structure of the collaboration network for department "A" is presented in Fig. 3(a). It is seen that the nodes are organized into roughly a star-structure. Specifically, there is a supper (central) node of degree $k = 22$ in the network, which contributes 61% of the total publications; while the degrees of the remaining (peripheral) nodes are less than $k = 6$. From the surveyed result, it is also learned that the super "node" is a senior researcher of broad research interest, with research directions covering almost all the directions of the peripheral "nodes". Besides the star-structure component, there also exists an isolated node-pair and two isolated nodes in the network. By checking the research background of the these staffs, it is revealed that the these staffs are working on a research directions not covered by the super "node".

Department "B" is a key laboratory of the Ministry of Education of China, which was founded in 2011 and now has 21 incumbent staffs. During the investigated period, $N_B = 17$ staffs have co-authored 39 papers in total, giving $L_B = 26$ collaboration links. The topological structure of the collaboration network for department "B" is presented in Fig. 3(b). It is seen that, different from department "A", the collaboration network of department "B" consists of 2 disconnected clusters, together with 5 isolated nodes. Comparing to department "A", it is seen that the node degrees of department "B" are more homogenously distributed, with k ranging from 1 to 7. By interviewing with staffs within department "B", it is learned that this is an interdisciplinary department founded by integrating three institutes from different departments. Although staffs in department "B" have the same research interest, they are actually working on different research directions and topics. Fig. 3(b) thus suggests that departmental affiliation is not the necessary condition for promoting scientific collaborations.

Department "C" is a provincial key laboratory, which was founded in 2005 and now has 36 incumbent staffs. During the investigated period, $N_C = 34$ staffs have co-authored 139 papers in total, giving $L_C = 54$ collaboration links. The topological structure of the collaboration network for department "C" is presented in Fig. 3(c). In comparison with the networks of departments A and B [Fig. 3(a) and (b)], the network of department C is featured by the existence of a few of tightly connected hub nodes. More specifically, the four largest-degree nodes [of node indices 14 ($k = 12$), 15 ($k = 9$), 17 ($k = 7$), and 22 ($k = 6$) in Fig. 3(c)] are globally connected and form the core of the community, to which most of the remaining nodes can reach by only one link. By interviewing the staffs, it is learned that this laboratory has long been established (back to 1980s) before promoting to the provincial key laboratory, and the research interest of the staffs are highly correlated. In particular, the four hub "nods" are of the same research direction, but working on different topics.

Our above analysis on the formation of collaboration patterns points to the fact that, among the many influenceable factors, research direction has the dominating role in affecting scientific collaborations. More specifically, the closer the research directions between two staffs is, the higher the possibility for them to collaborate with each other will be. This is evidenced in the collaboration patterns of department "A" [Fig. 3(a), where the research directions of the peripheral "nodes" are different from each other, but are all overlapped with the super "node"], department "B" [Fig. 3(b), where the research directions for nodes within the same cluster is highly centralized, but are different among the clusters], and department "C" [Fig. 3(c), where all "nodes" are working on different topics of the same research direction].

To explore further the crucial role of research direction on collaboration, we finally explore the collaboration patterns at the department level, i.e., combining nodes of the same color into a new

IOP Conf. Series: Journal of Physics: Conf. Series **1113** (2018) 012016 doi:10.1088/1742-6596/1113/1/012016

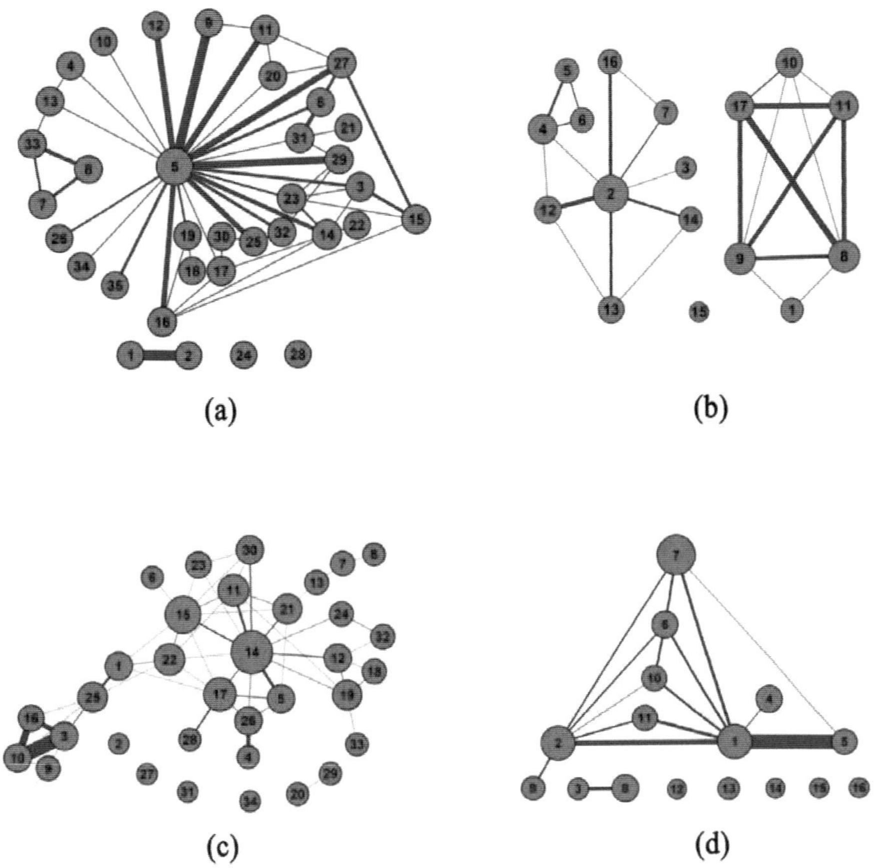

Figure 3. (Color online) (a) The collaboration network of department "A", which consists of $N_A = 35$ nodes and $L_A = 54$ links. The majority of the nodes are contained in the star-structure. (b) The collaboration network of department "B", which contains $N_B = 17$ nodes and $L_B = 26$ links. The network consists of 2 disconnected clusters. (c) The collaboration network of department "C", which contains $N_C = 34$ nodes and $L_B = 54$ links. The four largest-degree nodes are globally connected. (d) The collaboration network at the department level, which consists of 16 nodes and 34 links.

node and integrating the links accordingly. In such a way, the original network is compressed into a small-size network of only $M = 16$ nodes and $L = 16$ weighted links. (For a better presentation, only links of weight larger than 5 are kept). The network structure is presented in Fig. 3(d). Several features are observed in this departmental collaboration network, which manifests from another aspect the importance of research direction on scientific collaborations. Firstly, despite the large number of recorded publications, the network is still not globally connected, implying the difficulty of interdisciplinary collaborations (resembling the feature of department "B"). Secondly, the hub (non-hub) nodes stand for departments of fundamental (modern) science, e.g., the largest-degree ($k_1 = 7$) and second largest-degree ($k_2 = 6$) nodes symbolize, respectively, the chemistry (node 1) and biology (node 2) departments. This feature is similar to that observed in department "A", where the super "node" is of broad research interest and the peripheral nodes are of focused research interest. Thirdly, the connection between nodes

1 (the chemistry department) and 5 (the department of material science) is extremely strong (totally 95 co-authored papers), despite the fact that they belong to different disciplines. By checking the history of these two departments, it is found that node 5 was separated from node 1 in 2011, and many staffs still continue their collaborations since then. This finding, together with the finding of CDC in Sec. II, confirms the fact that scientific collaborations are influenced but not governed by the department affiliations.

4. Discussion and conclusion

The present work is featured by considering additional information about the network nodes and links. While collaboration networks have been extensively investigated in literature, the existing studies are mostly focusing on the network topological properties (e.g., degree distribution, small-world effect, and community structures) and less attention has been paid to the detailed information of the network nodes and links (e.g., affiliation, seniority, research direction, and acquaintanceship). In the present work, we pick up these missed information by the conventional methods used in social science (i.e., survey and interview), and revisit the assembling mechanism of collaboration patterns. With the additional information, we are able to analyze individually the factors that affect collaboration and, more importantly, identify the key elements, providing therefore a deeper understanding on the mechanism of scientific collaboration. We would like to note that all the findings are based on the specific publication dataset of SNNU, and it remains unknown to us whether the same results stand for other universities and organizations. In our further works, we shall keep on updating the SNNU publication dataset (e.g., extending the publication period), and, in the meantime, try to justify our findings by more datasets from other universities national- and world-wide. It is our hope that the findings, and also the new approach of collaboration analysis proposed in the present work, would draw attention from researchers in the relevant fields, and, by "collaborating" with each other, we together could unveil the origin of scientific collaboration inside universities. (All data used in the present work can be obtained from the authors upon request for research purposes.)

Summarizing up, based on a 6-year publication dataset of SNNU, we have constructed the scientific collaboration network of the university, and, combining with the staff additional information obtained through survey and interviews, explored in detail the assembly mechanism of various collaboration patterns. It is found that, among the many factors influencing scientific collaboration, the closeness of the research directions has a clear advantage over the others. The finding sheds new light on the mechanism of scientific collaborations, and might helpful to the policymaking and management of research activities in universities.

Acknowledgments

This work was supported by the National Natural Science Foundation of China under Grant No. 11375109 and by the Fundamental Research Funds for the Central Universities under Grant No. GK201601001.

References

[1] Guimerà R, Uzzi, B, Spiro J and Nunes Amaral L A 2005 *Science* **308** 697.
[2] Wuchty S, Jones B F and Uzzi B (2007) *Science* **316** 1036.
[3] Sonnenwald D H (2007) *Annual Review of Information Science and Technology* **41** 643.
[4] Newman M E J (2001) *Proc. Natl Acad. Sci. USA* **98** 404.
[5] Newman M E J (2001) *Phys. Rev. E* **64** 016132.
[6] Li M, Wu J, Wang D, Zhou T, Di Z and Fan Y (2007) *Physica A* **375** 355.
[7] Hsu J W and Huang D W (2009) *Phys. Rev. E* **80** 057101.
[8] Lee D, Goh K I, Kahng B and Kim D (2010) *Phys. Rev. E* **82** 026112.
[9] Zou S R, Xu Y Y, Xu X L, Wang J and He D R (2011) *Chinese Physics Letters* **28** 018902.
[10] Durugbo C, Hutabarat W, Tiwari A and Alcock J R (2011) *Information Sciences* **181** 3143.

[11] Martin T, Ball B, Karrer B and Newman M E J (2013) *Phys. Rev. E* **88** 012814.
[12] Uddin S, Hossain L and Rasmussen K (2013) *PLoS ONE* **8** e57546.
[13] Ke Q and Ahn Y Y (2014) *Phys. Rev. E* **90** 032804.
[14] Milojevic S (2014) *Proc. Natl Acad. Sci. USA* **111** 3984.
[15] Sun X, Lin H, Xu K and Ding K (2015) *Scientometrics* **104** 43.
[16] Liu J, Li Y, Ruan Z, Fu G, Chen X, Sadiq R and Deng Y (2015) *Physica A* **419** 29.
[17] Zhou J, Zeng A, Fan Y and Di Z (2015) *Scientometrics* **106** 805.
[18] Shen Z, Yang L, Pei J, Li M, Wu C, Bao J, Wei T, Di Z, Rousseau R and Wu J (2016) *Journal of Informetrics* **10** 82.
[19] Albert R and Barabasi A L (2002) *Rev. Mod. Phys.* **74** 47.
[20] Newman M E J (2003) *Society for Industrial and Applied Mathematics Review* **45** 167.
[21] Fang J Q, Bi Q and Li Y (2007) *Front. Phys.* **2** 109.
[22] Barrat A, Barthelemy M, Pastor-Satorras R and Vespignani A (2004) *Proc. Natl Acad. Sci. USA* **101** 3747.
[23] Newman M E J (2002) *Phys. Rev. Lett.* **89** 208701.
[24] Liu X F, Xu X K, Small M and Tse C K (2011) *PLoS ONE* **6** e26271.
[25] Cainelli G, Maggioni M A, Uberti T E and Felice A D (2015) *Scientometrics* **102** 673.
[26] Pepe A (2011) *Journal of the American Society for Information Science and Technology*, **62** 2121.
[27] Wasserman S and Faust K (1994) *Social network analysis: Methods and applications* (Cambridge University Press, Cambridge, England)
[28] http://www.fenqubiao.com/
[29] Catanzaro M, Caldarelli G and Pietronero L (2004). *Phys. Rev. E* **70** 037101.
[30] Girvan M and Newman M E J (2002) *Proc. Natl Acad. Sci. USA* **99** 7821.

Nonlinear cross-correlation among worldwide indexes

Min Lin[1], Longfeng Zhao[2], Yanjin Li[3]

[1] School of economics and management, Sichuan Normal University, Chengdu 61010, China
[2] Key Laboratory of Quark and Lepton Physics (MOE) and Institute of Particle Physics, Central China Normal University, Wuhan 430079, China
[3]College of International Studies, Sichuan University, Chengdu 610074, China

E-mail: joylinlee@vip.qq.com (Y. Li)

Abstract. The q-dependent cross-correlation between worldwide stock indexes have been analyzed. The correlations among different fluctuations have different identities. The correlation-based network has been constructed for different magnitude of fluctuations. We found apparent community structure related to geographical location. Those results have deepened our understanding about the cross-correlation between stock markets.

1. Introduction

The collective behaviors of cross-correlations between different financial assets has become extremely attractive [1–3]. The traditional cross-correlation analysis always relied on some linear tools such as Pearson cross-correlation coefficient. But actual linear correlations are very rare in real world situation. In order to take into account the non-linearity and non-stationarity in real-world data, detrendlized methods have been proposed to tackle the non-linear and non-stationary of real world datasets.

The most popular nonlinear and non-stationary time series analysis method is the detrended fluctuation analysis(DFA) [4]. The generalization of DFA method named detrended cross-correlation fluctuation analysis (DCCA) has been proposed to quantify non-linear cross-correlations between a pair of non-stationary signals [5]. DFA and DCCA are then extended to their multifractal versions: MFDFA and MFDCCA, respectively [6–8]. Those tools have been widely used in a broad range of systems including biological, financial to physical systems [9–17]. Recently a cross-correlation coefficient similar to the Pearson coefficient, which is based on DCCA, the detrended cross-correlation coefficient $\rho(s)$, was introduced in ref. [18]. This coefficient has been applied to non-stationary signals and quantifies the correlations among fluctuations of detrended non-stationary signals at a given detrending scale s [19].

The DCCA coefficient $\rho(s)$ has been widely employed to study the non-linear cross-correlations among financial time series [10, 20–23]. A more recent extension of the DCCA coefficient $\rho(s)$ based on the q-dependent fluctuation function F_q from MFDFA and MFDCCA [6, 8, 24] has been proposed. Kwapień et al. [24] recently indicated that this method could be employed to analyze the empirical data from physical, biological, social and financial systems. Our focus here is on worldwide stock indexes.

Here the q-dependent cross-correlation coefficient is used to quantify the cross-correlations among the return time series of worldwide stocks indexes. Although the q-dependent cross-correlation among stock return time series have been analyzed in a very detailed manner for both high frequency and daily data [16, 25]. But the properties of nonlinear cross-correlation among worldwide indexes is still unclear. Here we calculate the q-dependent cross-correlation matrices $\mathbf{C}(q, s)$. Then the statistical properties of the matrices at different multifractal orders q and varying time scales s have been calculated. We then use the planar maximally filtered graph (PMFG) method [26] to characterize the cross-correlation matrices as complex networks and to analyze their basic topological features. The PMFG network has been widely used in the research about the cross-correlation matrices [17, 25, 28]. We use the community detection algorithm to detected the community structure of the PMFG networks [29–31]. The PMFG networks at different multifractal order q and detrended scales s have unique structures and related with very clear geography identities. The paper is organized as follows. In Sec. 2 we introduce the data and methodologies used in this paper. In Sec. 3 we present the main empirical results. The last section provides our conclusion .

2. Data and Methodology

2.1. Data

The dataset we use include 37 indexes all over the world. There are 5 indexes from the US, 15 indexes from the European, 13 indexes from Asia, 5 indexes from America and 1 index from Africa. Every index includes 3381 data points from 2003-01-03 to 2015-12-18.

2.2. q-dependent cross-correlation analysis

We can be obtained the q-dependent cross-correlation coefficient from the following procedure [8]: (i) For a pair of time series x_i and y_i, $i = 1 \dots l$. The integrated time series are generated first

$$\chi^x(k) = \sum_{i=1}^{k} x_i - \langle x \rangle, k = 1 \dots l, \tag{1}$$

$$\chi^y(k) = \sum_{i=1}^{k} y_i - \langle y \rangle, k = 1 \dots l. \tag{2}$$

(ii) Then $\chi^x(k)$ and $\chi^y(k)$ are divide into $2M_s = 2 \times \text{int}(l/s)$ non-overlapping boxes of length s from the beginning and the end of two integrated time series$\chi^x(k)$ and $\chi^y(k)$. The local trends for each segment $v(v = 1, \dots, 2M_s)$ can be estimated by a least-square fit and are subtracted from $\chi^x(k)$ and $\chi^y(k)$ to detrend the integrated series. Then the residual signals X, Y equal to the differences between the integrated signals and the mth-order polynomials $P_{s,v}^{(m)}$ fitted to these signals are as follows:

$$X_s(i, v) = \sum_{i=1}^{s} \chi^x(vs + i) - P_{X,s}^{(m)}(i, v), \tag{3}$$

$$Y_s(i, v) = \sum_{i=1}^{s} \chi^y(vs + i) - P_{Y,s}^{(m)}(i, v). \tag{4}$$

The covariance and variance of X and Y in a box v are defined as:

$$f_{XY;s}^2(v) = \frac{1}{s} \sum_{i=1}^{s} X_s(i, v) Y_s(i, v), \tag{5}$$

$$f_{ZZ;s}^2(v) = \frac{1}{s} \sum_{i=1}^{s} Z_s^2(i, v), \tag{6}$$

here Z represents either X or Y.
(iii) We then define the fluctuation functions at multifractal order q and detrending scale s as:

$$F^q_{XY}(s) = \frac{1}{2M_s}\sum_{v=1}^{2M_s} sgn[f^2_{XY;s}(v)]|f^2_{XY;s}(v)|^{q/2}, \quad (7)$$

$$F^q_{ZZ}(s) = \frac{1}{2M_s}\sum_{v=1}^{2M_s}[f^2_{ZZ;s}(v)]^{q/2}. \quad (8)$$

Then the q-dependent cross-correlation coefficient between two time series x_i and y_i is defined as:

$$\rho(q,s) = \frac{F^q_{XY}(s)}{\sqrt{F^q_{XX}(s)F^q_{YY}(s)}}. \quad (9)$$

When $q = 2$ we have the original detrended cross-correlation coefficient $\rho(s)$ [18].

The q-dependent cross-correlation coefficient is bounded in the range of $[-1, 1]$ when $q \geq 0$. This coefficient has arbitrary value when $q < 0$ [24].We only use the case when $q > 0$. The exponent q acts as a filter. When $q > 2$ the boxes with large fluctuations contribute to $\rho(q, s)$ the most, but when $q < 2$ the boxes with relatively small values will dominate the fluctuation function. Meanwhile, the detrend scale s represents the different length of trend in the time series.

2.3. Planar maximally filtered graph

Due to the dense representation of the cross-correlation matrix among stock indexes, its very hard to distinguish the structural information from the random noise. Thus we need a sparse representation of the cross-correlation matrix to illustrate the structural information. The complex network approach has been widely used to analyze the cross-correlation matrix due to its straightforward interpretation. Here the planar maximally filtered graph (PMFG) method [26, 27] is used to construct networks based on cross-correlation matrices $\mathbf{C}(q, s)$. The PMFG algorithm converts the dense correlation matrix into a sparse representation q-PMFGs.

As it is known from ref. [26], PMFG keeps both the hierarchical organizations of the minimum spanning trees(MST) and it also induces cliques, especially those three and four nodes cliques. The basic topological parameters such as clustering coefficient C, the shortest-path length L and the assortativity A [32] have also been calculated. A heterogeneity index γ is also used to measure the heterogeneity of the network [33]:

$$\gamma = \frac{N - 2\sum_{ij\in\{e\}}(k_i k_j)^{-1/2}}{N - 2\sqrt{N-1}}. \quad (10)$$

Here k_i and k_j are the degrees of nodes i and j connected by edge $\{e_{ij}\}$.

3. Results

Here in figure 1 we present the basic statistics of the q-dependent cross-correlation matrices. We can see that there is a very obvious transition for the first four moments. The mean correlation for small multifractal order q is relatively larger than those for higher order q. The shapes of the distributions change from left skew to right ones. Obvious distribution transition gives a hint about the different cross-correlation structures for small and large fluctuations of the world-wide indexes.

Figure 2 gives the q-PMFG networks for different multifractal order q and detrending scales s. A very interesting phenomena is that those q-PMFG networks carry obvious geography

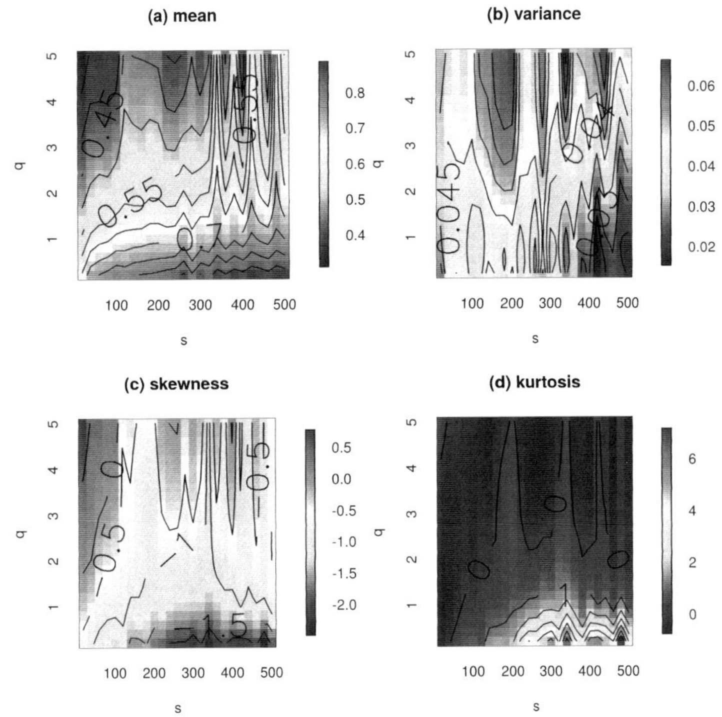

Figure 1. The first four moments of the q-dependent cross-correlation matrices for each multifractal order q and detrending scale s.

information. Those stock indexes belong to the same geography location tend to reside in the same cluster. We use the infomap community detection algorithm to detect the community structures of the q-PMFGs [29]. The indexes in the same community are placed in the same circle. Different colors are related to the geography information of the indexes. We find that for large q the indexes are mixed with each other and for small q the community structures are more geography related and separated from each other.

Here in figure 3, we present the basic topology quantities for q-PMFG networks. The clustering coefficient C for small multifractal order q and small detrending scale s is relative small which means the small fluctuations are strong correlated with each other. The heterogeneous index also revealed the heterogeneous network structure for small q and s.

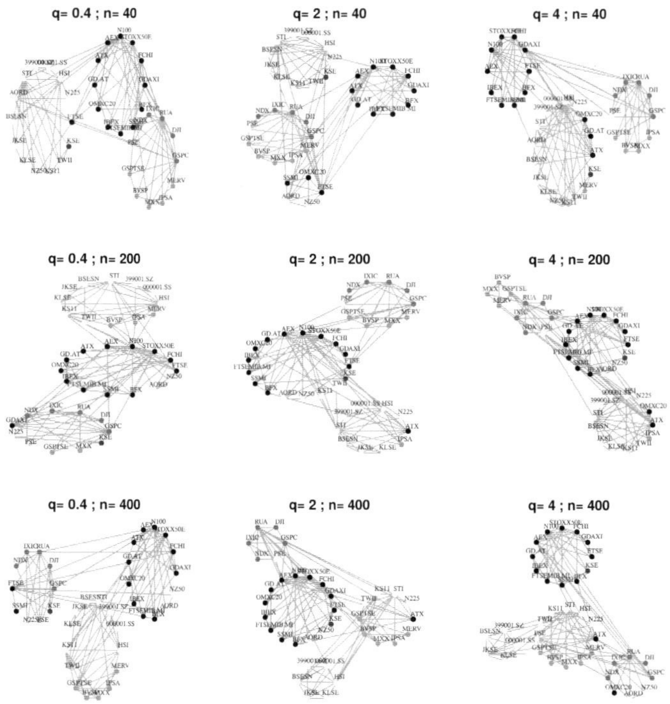

Figure 2. The q-PMFG networks for different multifractal orders q and detrending scales s. The yellow nodes are those indexes from Asian. The black nodes are those indexes from European. The red ones are those indexes from north America. The sky blue nodes are those indexes from South America. The only blue node is the index from Africa. The location of each node in the network is determined by the community detection algorithm. Each circle in the network represents one community.

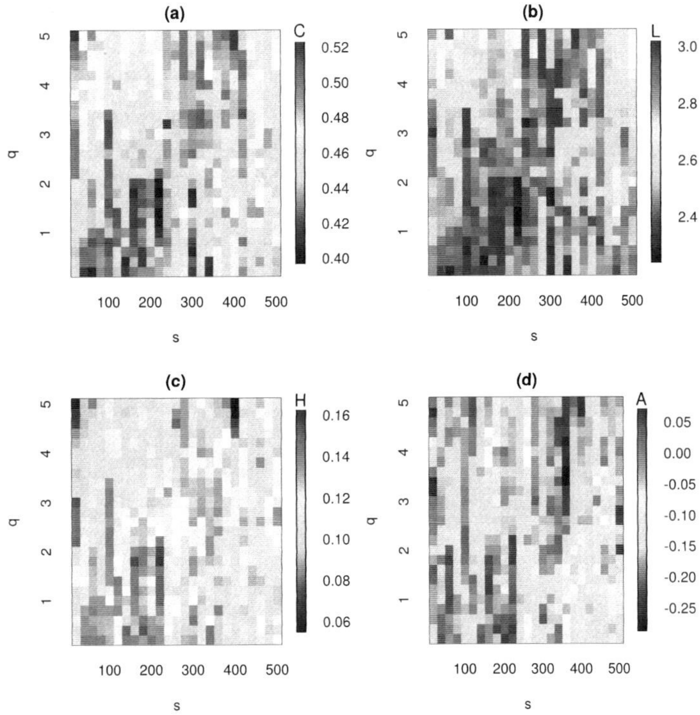

Figure 3. The basic topology quantities for q-PMFG at different multifractal orders q and detrending scales s. C is the clustering coefficient, L is the shortest path length, H is the heterogeneity index and A is the assortative of the network.

4. Conclusion

In conclusion, the detrended cross-correlation coefficient has been used to analyze the worldwide stock indexes. The fluctuations at different magnitudes have their own identities. The cross-correlation among small fluctuations are stronger than those among large fluctuations. The q-PMFG networks have also been employed to analyze the correlation structures of the q-dependent cross-correlation matrices. The q-PMFGs give a very clear geography related community structures. Whilst the community structures at different fluctuation orders are quite different from each other. The results from this paper have deepened our understanding about the correlation structure among worldwide stock markets.

Acknowledgement

L. Zhao is supported in part by the Programme of Introducing Talents of Discipline to Universities under grant NO. B08033 and the program of China Scholarship Council (No. 201606770023).

References

[1] Plerou V, Gopikrishnan P, Rosenow B, Amaral L A N and Stanley H E 1999 *Physical Review Letters* **83** 1471–1474

[2] Laloux L, Cizeau P, Bouchaud J P and Potters M 1999 *Physical Review Letters* **83** 1467–1470

[3] Kwapień J and Drożdż S 2012 *Physics Reports* **515** 115–226

[4] Peng C K, Buldyrev S V, Havlin S, Simons M, Stanley H E and Goldberger a L 1994 *Physical Review E* **49** 1685–1689

[5] Podobnik B and Stanley H 2008 *Physical Review Letters* **100** 084102

[6] Kantelhardt J W, Zschiegner S A, Koscielny-Bunde E, Havlin S, Bunde A and Stanley H E 2002 *Physica A: Statistical Mechanics and its Applications* **316** 87–114

[7] Zhou W X 2008 *Physical Review E* **77** 066211

[8] Oświecimka P, Drożdż S, Forczek M, Jadach S and Kwapień J 2014 *Physical Review E* **89** 023305

[9] Alvarez-Ramirez J, Rodriguez E and Echeverria J C 2009 *Physical review. E, Statistical, nonlinear, and soft matter physics* **79** 057202

[10] Podobnik B, Jiang Z Q, Zhou W X and Stanley H E 2011 *Physical Review E - Statistical, Nonlinear, and Soft Matter Physics* **84** 1–8

[11] Stan C, Cristescu M T, Luiza B I and Cristescu C P 2013 *Journal of Theoretical Biology* **321** 54–62

[12] Rak R, Drozdz S, Kwapien J and Oswiecimka P 2015 *EPL (Europhysics Letters)* **112** 48001

[13] Kristoufek L 2014 *Physical Review E* **022802** 1–5

[14] OświÉcimka P, Livi L and Drozdz S 2016 *Physical Review E* **94** 42307

[15] Salat H, Murcio R and Arcaute E 2016 *Physica A* **473** 1–14

[16] Kwapień J, Oświcimka P, Forczek M and Drozdz S 2017 *Physical Review E* **95** 1–10

[17] Zhao L, Li W, Yang C, Han J, Su Z and Zou Y 2017 *PLoS ONE* **12** 1–23

[18] Zebende G 2011 *Physica A: Statistical Mechanics and its Applications* **390** 614–618

[19] Kristoufek L 2014 *Physica A: Statistical Mechanics and its Applications* **402** 291–298

[20] Wang G J, Xie C, Chen S, Yang J J and Yang M Y 2013 *Physica A: Statistical Mechanics and its Applications* **392** 3715–3730

[21] Wang G J, Xie C, Chen Y J and Chen S 2013 *Entropy* **15** 1643–1662

[22] Zebende G F, Da Silva M F and MacHado Filho A 2013 *Physica A: Statistical Mechanics and its Applications* **392** 1756–1761

[23] Sun X and Liu Z 2016 *Physica A: Statistical Mechanics and its Applications* **444** 667–679

[24] Kwapień J, Oswięcimka P and Drozdz S 2015 *Physical Review E - Statistical, Nonlinear, and Soft Matter Physics* **92** 1–11

[25] Zhao L, Li W, Fenu A, Podobnik B, Wang Y and Stanley H E 2018 *Journal of Statistical Mechanics: Theory and Experiment* **2018** 23402

[26] Tumminello M, Aste T, Matteo T D and Mantegna R N 2005 *Proceedings of the National Academy of Sciences of the United States of America* **102** 10421–10426

[27] Zhao L, Wang G J, Wang M, Bao W, Li W and Stanley H E 2018 *Physica A, https://doi.org/10.1016/j.physa.2018.05.039.*

[28] Zhao L, Li W and Cai X 2016 *Physics Letters A* **380** 654–666

IOP Conf. Series: Journal of Physics: Conf. Series **1113** (2018) 012017 doi:10.1088/1742-6596/1113/1/012017

[29] Rosvall M and Bergstrom C T 2008 *Proceedings of the National Academy of Sciences of the United States of America* **105** 1118–23
[30] Han J, Li W, Su Z, Zhao L and Deng W 2016 *The European Physical Journal B* **89** 272
[31] Han J, Li W, Zhao L, Su Z, Zou Y and Deng W 2017 *PLoS ONE* **12** e0188655
[32] Boccaletti S, Latora V, Moreno Y, Chavez M and Hwang D U 2006 *Physics Reports* **424** 175–308
[33] Estrada E 2010 *Physical Review E* **82** 66102

Measurement of phonetic differences in Guangxi dialects

Ruyue Sun[1], **Jiawei Chen**[2]

[1] School of Ethnic Culture, Yunnan Minzu University, Kunming, 650031, China
[2] School of Systems Science, Beijing Normal University, Beijing, 100875, China

E-mail: chenjiawei@bnu.edu.cn (Jiawei Chen)

Abstract. The study of dialect differences plays an important role in the research of linguistics. In the recent years there has been a trend to measure dialect differences with computational models. From the perspective of phonology, a syllables of a Chinese character can be divided into three parts: The Initials, the Finals and the Tone. The Initials are usually composed of consonants, and the Finals are composed of vowels or vowels and consonants. Therefore, as long as the quantitative characteristic of vowels and consonants are described, and the Five-Degree markings method of tones are used, the differences between Chinese syllables can be calculated. We developed a model and applied it to the measure and analysis the differences of dialects in Guangxi. The analysis results accord with empirical data of Guangxi province dialectal subdivisions and this shows the validity of the model.

1. Introduction

Linguistics have been studying dialect differences for a long time[1, 2, 3, 4, 5]. The research can not only support subdividing regions according to dialects [6], but also provide a special perspective for studies on social phenomena such as cultural differences, population migration, and integration of ethnicities [7]. The differences between dialects are prominent in the aspects of phonetic, while the differences in vocabulary and grammar are relatively small [8]. Therefore, the measurement of phonetic differences becomes an important part of dialect research.

The differences of phonetics of Chinese dialect are reflected in their Initials consonant, Finals vowels and Tone, on which many scholars study this problem. Lu Zhiji [9], Yang Dingfu [10] and Xie Jianyou [11] calculated the dialect differences by counting the number of vowels and consonants found in different dialects. These works have laid the foundation for the study of Chinese phonetic differences, but due to the limited technical methods at that time, these studies only made rough descriptions of phonetic differences among dialects, only presenting whether or not the features of each phonetic have appeared. This paper will quantitatively describe the phonetic features of each phoneme based the work above, characterize it use a 3 or 4 dimension vector, and assign the weights of each part to calculate the distance of dialects among different regions.

In respect of the research region, we choose Guangxi Zhuang Autonomous Region (Guangxi Province), which has the largest minority population in China. In Guangxi province, there are 12 ethnicities settle down in the district such as Zhuang, Han, Yao and Miao, etc. In such a social context Guangxi has developed a large number of Chinese dialects nowadays [12]. According to "Language Atlas of China" [6], modern Chinese dialects are divided into ten dialectal areas such as northern dialect, Hakka dialect, Wu dialect, and Cantonese dialect, while Guangxi province includes the southwest Mandarin dialect, Cantonese dialect, Xiang dialect, Hakka dialect, Min dialect, Ping dialect six categories, even some unknown native languages. Therefore, Guangxi province are called "the mine of languages" by many people, and this shows great research value [11].

2. The Model

Traditional phonology divides a Chinese syllable into three parts: initials consonants, Finals vowels and tone, so the differences of Chinese pronunciation among regions are reflected in these three parts. We characterize these parts respectively.

2.1. Initials Consonants

In most cases, the Initials are consonants at the beginning of the syllables. There are also some syllables without consonants, which are called zero consonants. Therefore, the initial consonants differ in the consonant pronunciation. Consonants are the sounds generated in the mouth or pharynx as the air flow is obstructed. The pronunciation features can be described from the four aspects. They are place of articulation, obstruction characteristics, airflow intensity and loudness.

The place of articulation refers to the part where the airflow is obstructed when pronouncing it. There are roughly 11 areas from the lips to the throat [13], each of which place is assigned a number with the sequence (starting from 1/11 and with 1/11 increment) accordingly. The place of articulation and the characteristic value are shown in Table 1.

Table 1. Place of articulations and characteristic value.

Place of articulation	Bilabial	Labiodental	Dental	Alveolar	Post-alveolar	Retroflex
characteristic value	0.091	0.182	0.273	0.364	0.455	0.545
Place of articulation	Palatal	Velar	Uvular	Pharyngeal	Glottal	
characteristic value	0.636	0.727	0.818	0.909	1	

The pronunciation method refers to the status of the throat, mouth, and nasal control mode when pronouncing the syllable. Nine categories can be divided into according to the extent of the obstruction[13]. The assigned values are shown in Table 2.

Table 2. Obstruction and its characteristic value.

Pronunciation methods	Plosive	Affricate	Nasal	Trill	Tap/Flap
characteristic value	0.111	0.222	0.333	0.444	0.556
Pronunciation methods	Lateral	Lateral Fricative	Fricative	Semivowel	
characteristic value	0.667	0.778	0.889	1	

According to whether there are air supply or airflow strength, two categories can be divided into. No air supply condition is assigned 0.333, and the existence of air supply is assigned 0.667, while the condition of possessing no trait should be assigned 1.

According to whether there exists the vibration of the vocal cords, voiceless and voiced consonants can be characterized. These two categories have significant differences in loudness. The voiceless consonant and the voiced one are assigned 0.33 and 0.66 respectively (the vowel is the loudest, so it should be assigned 1).

For any consonant, four values can be used to characterize their pronunciation features. For example, the phoneme /m/ can be described as the vector (0.091, 0.333, 1, 0.667). For zero initials, the four dimensions should be set to 0.

2.2. Finals Vowels

The Finals refers to the vowel part usually at the end of a syllable or the combination of the vowel and consonants, and it can be divided into rhyme head, rhyme body and rhyme tail. The rhyme head and the rhyme body are composed of vowel, while the rhyme tail is made up of vowels or consonants. Therefore, as long as we characterize the vowel pronunciation features and combine with the consonant pronunciation features discussed in the previous section, we can characterize the differences of Finals vowels. Vowels are unobstructed pronunciation, but they are still restrained by airflow during pronunciation. Their vocal characteristics are mainly investigated from three aspects: the position of the tongue (the upper/lower side, the front/back side in the mouth), and the roundness of the lip.

According to the position of the tongue during pronunciation, 7 levels of the vertical height are categorized from the most upper side to the lowest side [14]. The heights position and their characteristic value are shown in Table 3.

Table 3. Tongue position and its characteristic value.

Tongue position	Highest	Higher	High	Middle	Low	Lower	Lowest
characteristic value	0.143	0.286	0.429	0.572	0.715	0.858	1

The location of the tip of the tongue can be divided into three categories: front vowel, middle vowel, back vowel [14]. Their characteristic values are shown in Table 4.

Table 4. Anterior and posterior tongue and its characteristic value.

Anterior and posterior tongue	Front	Middle	Back
characteristic value	0.333	0.666	1

The differences of the lip roundness cause the different shapes of the resonator, forming a variety of vowel-specific sounds. The degree of the roundness of the lip can be divided into 7 types [14], the characteristic values of which are shown in Table 5.

Table 5. Lip roundness and its characteristic value.

Lip roundness	Open	Near-Open	Open-Mid	Mid
characteristic value	0.143	0.286	0.429	0.572
Lip roundness	Close-Mid	Near-Close	Close	
characteristic value	0.715	0.858	1	

The phonemes used as rhyme head and rhyme body are vowels, and they can be characterized by the above three features. The rhyme tail may be vowel or consonant, which can be characterized by seven features, and the first three parts are used to describe the vowel, and the last four are used to describe the consonant. Therefore, we need a $3 + 3 + 7 = 13$ dimension vector to describe a complete Finals vowel. For Finals lacking rhyme head or rhyme tail, the lacking parts should be set to zero. For example, /uan/. The rhyme head is /u/ would be (0.143, 1, 1), the rhyme body /a/ would be (1,0.333,0.286), and the rhyme tail /n/ would be (0.364,0.333,1,0.667,0,0,0). Another example is /iau/, the rhyme head /i/ would be (0.143,0.333,0.143), the rhyme body /a/ would be (1,0.333,0.286), and the rhyme tail /u/ would be (0,0,0,0,0.143,1,1).

2.3. Tone
The tone refers to the super segmental phoneme attached to a syllable, mainly shows the pitch of a syllables. The tone system of modern Chinese dialects is more complicated than that of Mandarin. Most areas south to the Yangtze River have more than five kinds of tones. For example, in Nanchang county of Jiangxi province there are 13 kinds of tones. Therefore, differences among tones can hardly be described using four tones in Mandarin. In this paper, Zhao Yuanren's Five-Scale Marking Method is deployed. This method uses two or three integers ranging from 1 to 5 to represent a tone. For the 2-digit tone, we insert the mean value of these two numbers into the middle of them, so as to unify all representations into 3-digit. The middle tone serves as the dividing line, and the tone is divided into two parts. In this case, a tone has five possible scenarios: flat, rise, fall, rise-fall and fall-rise. In order to describe the pitch and tone characteristics, we use the tone 3-digit representation and the slopes of two tone sections to characterize a tone. Considering all five values range from 1 to 5, We divide this 5-dimensional vector by 5 so that the characteristic values of both the initial consonant and the final vowels are between 0 and 1.

For example, if a certain tone representation is [1, 5], it should be regularized as 3-digit format as (1, 3, 5). Both the slope of the first segment and the second segment are 2. The five-dimension representation should be (1, 3, 5, 2, 2). After regularization and normalization, (0.2, 0.6, 1, 0.4, 0.4) can be obtained. Another example is a tone value of (2, 4, 3), its final five-dimension representation is (0.4, 0.8, 0.6, 0.4, -0.2).

2.4. Measuring the phonological differences
The Euclidean Distance is used to calculate the differences of initial consonant, finals vowels, and tone in different regions respectively. For example, the initial consonant of the Chinese character C may have different pronunciation in different places, but all of them can be described as a four-dimension vector based on their phonetics. Calculating the Euclidean distance between these four-dimension vectors, a distance matrix D_C can be obtained, where $D_C(i,j) = \sqrt{\sum_{k=1}^{4} (S_{C,ik} - S_{C,jk})^2}$, where i, j are the region ID, while k is the feature ID. Selecting the initial consonants of m Chinese characters, and the distance matrices $D_1, D_2, ..., D_m$ can be obtained. Differences of the initial consonants in different regions i and j can be calculated by $D(i,j) = \sum_{n=1}^{m} D_n(i,j)/m$. The calculations in finals vowel and tone are similar to initial consonants. Given a certain weight to initials consonants, finals vowels and tones, we can get the total difference between all regions.

3. Data and Results
3.1. Data Source
The data of this paper comes from "Linguistic Atlas of Chinese Dialects (Phonetics)" [15]. Initials consonants of 71 Chinese characters, finals vowels of 101 Chinese characters and tones of 9 Chinese characters are collected.

There are 930 dialect points were set in the book, and 66 of which locate in Guangxi province. There are 111 county-level regions in Guangxi province, and they are categorized into the 66 dialect points according to the exciting dialect division.

3.2. Results
Following the above instructions, distance matrices of initial consonants, vowels, and tones are calculated between 66 dialect points. Hierarchical clustering results of the three distance matrices are shown as Figure 1. We divided all the counties into 7 groups based on the clustering

results. The same color in the cluster plot and map represents the same category. Because the classification results of different clustering plots are inconsistent, it is inconvenient to make comparisons.

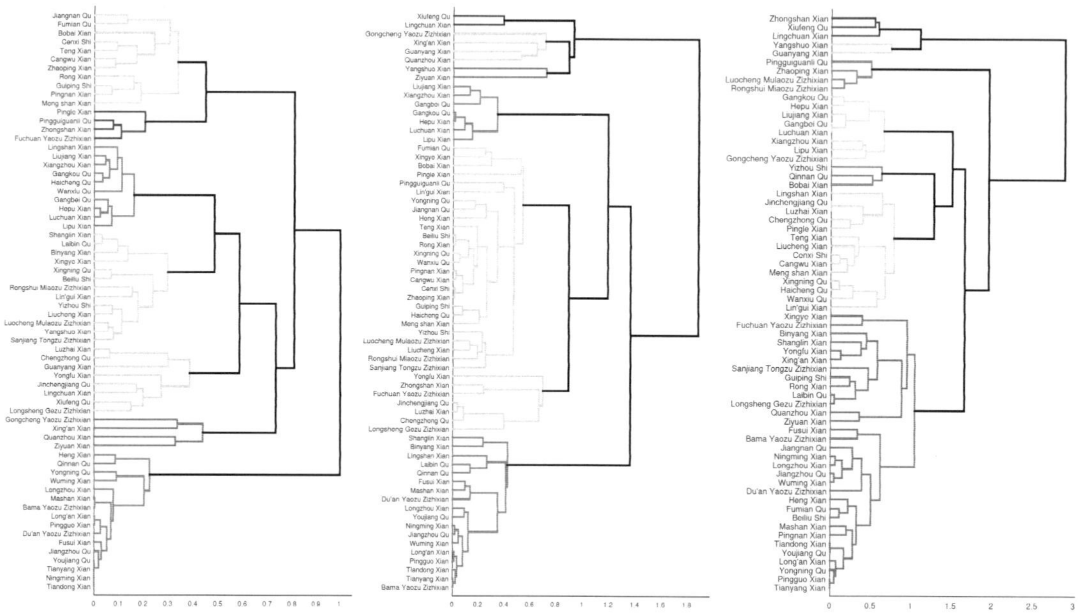

Figure 1. Initials consonant, Finals vowels and Tones Clusters of 66 Representative Points in Guangxi province.

It is clearly shown that data has been divided into several categories in all three figures, indicating that the differences between Guangxi dialects are significant. It is consistent with the truth that Guangxi province has a wide range of dialects.

the 111 county-level regions which are represented by the 66 dialect points are projected on the map according to the clustering results, and we can clearly see the aggregate situation of different dialects. It is shown in Figure 2.

Figure 2. Initials consonant, Finals vowels and Tones geographic division of 111 counties in Guangxi province.

As can be seen from Figure 2, the clustering patterns of the initials consonants and the vowels are similar. However, the geographic pattern of tones distribution is scattered and has no obvious pattern in this region, except the southwest Mandarin dialect.

IOP Conf. Series: Journal of Physics: Conf. Series **1113** (2018) 012018 doi:10.1088/1742-6596/1113/1/012018

The Final is more important than the initial and the tone, but the vector which denotes the final is larger than others. So we set the same weights for the three parts of a syllables, and the distance matrices of total difference between regions can be calculated. Figure 3 shows the clustering result and the map.

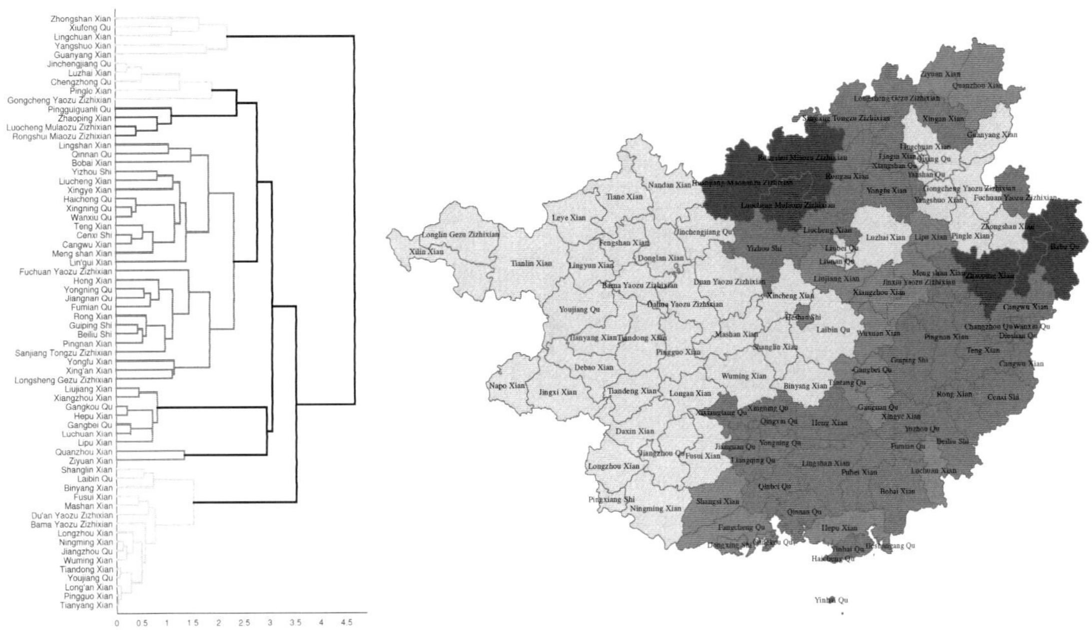

Figure 3. The clustering results and geographic distribution of total difference between regions of Guangxi province.

The clustering results of the total difference are similar to those of the finals vowels. The more obvious ones are the southwest Mandarin dialect, the northeast Xiang dialect, the east Cantonese region, and the middle region mainly popularized Mandarin and few Cantonese. The clustering results significantly distinguish the dialect area. As for the Guiliu Mandarin regions, the model also distinguishes two to three categories, and it is useful for further studying in dialectal subdivisions.

4. Conclusion and Discuss

The results the clustering analysis of dialect differences in different areas of Guangxi province accord with the existing dialect subdivisions, and this shows the validates the model. Also, the model can also provide a basis for dialect division.

Due to the difficulty of data obtaining and limited samples are selected, therefore the results are not precise. The clustering results still are not exactly the same as the existing partition. In addition, there are imperfections in the model when describing dialects in some areas. For example, in some dialects there are double vowels as rhyme body and double consonants as the initial consonants. This model can only deal with these problems in a simplified way, we mainly use the primary consonant or vowel and adjusting it according with the second consonant or vowel. We will improve these two deficiencies in the future work.

Acknowledgement

This work was supported by the National Social Science Fund of China under Grant No. $14BSH024$.

References

[1] Dong Shaoke, On the Differences of Word Formation in the Vocabulary of Chinese Dialects, Journal of Shangdong Teachers University (Humanities and Social Sciences), 2002, 47(1):39-43.
[2] Xiao Li-ming, On the Internal Discrepancy and the Contributing Factor of the Localism in Sinology Language in Tongren Prefecture, Journal of Tongren University, 2007, 1(4):64-68.
[3] Li Xin,The Comparisons of Yangxian Dialect in Qinba Area and Putonghua,Sichuan University of Arts and Science Journal. 2017, 27(4)94-98.
[4] Ji Yun-qi, The Variation between South and North of Chinese Dialects and Their Cultural Elucidation. Qilu Journal, 2009, 190:158-160.
[5] Xu Yue, The Internal Phonological Differences of Hangzhou Dialect in Zhejiang Province, Dialect. 2007,1:10-14.
[6] Institute of Linguistics, Chinese Academy of Social Sciences, Language Atlas of China (The Second Edition). The Commercial Press, 2012.
[7] Zhou Zhenhe, The Migration Trends of Chinese and Their Influence on the Geography of Chinese Dialects. Human Geography, 1988(1):57-64.
[8] Ye Mansheng, Xu Tongqiang. Introduction to Linguistics (The Fourth Edition). Peking University Press. 2010.
[9] Lu Zhiji, Measurement and Description of the Relationship between the Chinese Dialects, Social Sciences in China, 1987(1):155-166.
[10] Yang Dingfu, Research on the Dialectal Division of Chinese Dialects by Metrology, Linguistic Researches, 1994(3):42-47.
[11] Xie Jianyou, Zhang Zong. A Comparative Study on the Artificial Statistics and Computer Computation in the Case of Classification /Distribution of Chinese Dialects in Guangxi Province. China Journal of Dialectology, 2014, (4):162-174.
[12] Luo Cai-juan, Relation History of Guangxi Nationalities: a History of National Integration. Journal of Guangxi Normal University for Nationalities, 2011, 28(4):9-12.
[13] Hu Zhuanglin, Linguistics tutorial (The Fourth Edition). Peking University Press. 2013.
[14] Xing Fuyi, Wu Zhenguo. Introduction to Linguistics (The Second Edition). Huazhong Normal University Press. 2015.
[15] Cao Zhiyun. Linguistic Atlas of Chinese Dialects (Phonetics). The Commercial Press, 2008.

IOP Conf. Series: Journal of Physics: Conf. Series **1113** (2018) 012019 doi:10.1088/1742-6596/1113/1/012019

Key nodes mining of root network and phoneme network of modern Chinese characters

Rong Huang[1], Quanyu Zou[2], Kaiyu Xu[2], Mingyang Cai[2], Wenjun Zhang[3], Weibing Deng[3] and Bo Liu[4]

[1] National Bases for Basic Scientific Research and Talents Cultivation, College of Physical Science and Technology, Central China Normal University, Wuhan 430079, China

[2] College of Physical Science and Technology, Central China Normal University, Wuhan 430079, China

[3] Key Laboratory of Quark and Lepton Physics (MOE) and Institute of Particle Physics, Central China Normal University, Wuhan 430079, China

[4] Engineering Training Center, School of Mechanical and Electronic Engineering, Wuhan University of Technology, Wuhan 430070, China

E-mail: wenjun@mails.ccnu.edu.cn and wdeng@mail.ccnu.edu.cn

Abstract. Roots and phonemes are two important ingredients (*but not* independent ones) when learning the Chinese language. In order to optimize the learning strategy, various key nodes mining algorithms are implemented both in the root network (RN) and phoneme network (PN) of modern Chinese characters. The lists of important roots and phonemes, and eventually the important characters are dug out and suggested. Rather than the commonly used frequency dictionary of characters, one would find that results from the key nodes mining are more systematic. Moreover, roots and phonemes are studied from perspective of the multi-layer network, in the interest of exploring the correlation between them, which are also very useful when learning the Chinese language.

1. Introduction

Considering human languages as the dynamical complex systems, quantitative linguistics, or mathematical linguistics [1, 2, 3] deals with the language statistics, language structure, language evolution, language learning, etc, by using the methods of probability theory [4, 5, 6, 7, 8], machine learning or deep learning [9], statistical physics [3], complex networks [2] and so on. These studies are useful for understanding the origin of human languages [10], the physical and mathematical laws behind the language phenomena [11], the self-adaptive mechanisms of languages [12], the reasons embedded in the evolution of languages [2], the optimization of information transfer [13], etc. They are also applicable for some concrete problems of natural language processing, such as text classification, authorship attribution, information searching, machine translation, etc [14, 15, 16, 17].

Complex networks are very powerful methods for studying human languages [2], relevant works have been rapidly and vastly carried out in the past decades (see the literatures collection website [18] of this direction-Physicists' papers on natural language from a complex systems viewpoint, 1989-2018). The researches in this area can be roughly divided into three categories:

1) Exploring the multi-level features of human languages: language system is a multi-level complex system, from different levels one can get different kinds of language networks [19, 20, 21]. For example, by using the same set of sentences of Chinese, the following three levels of networks can be constructed: (I) Chinese character co-occurrence network; (II) Chinese syntactic network; (III) Semantic role network. The nodes in (I) are characters, while the links are formed between adjacent characters in a sentence, it can be used to study the formation mechanism of Chinese vocabulary. The nodes in (II) are words, and the links are determined by the syntactic functional relationship. Network III is constructed by a set of sentences that have been tagged with semantic roles, namely the Chinese semantic network. Cancho and Solé [22], Motter and Moura [23], Liu and Cong [19] studied and compared these three kinds of networks, analyzed the different cognitive mechanisms of information expression [24].

2) Using the topological quantities of networks to study the linguistic typology: linguistic typology not only studies the classification of languages, but also pay much attention to find the common features of human languages. Liu et al. constructed the syntactic networks based on two different languages, and found that syntactic networks with the same diameter would have obvious different average degree, the power law exponent of degree distribution, the clustering coefficient, the average shortest path, and the centrality [25]. Later on, they constructed 15 syntactic networks of different languages, and compared the topological parameters of different networks to classify different languages [24]. Amancio et al. extended the methods of linguistic classification and text categorization based on topological parameters of complex networks to the dynasty of literary works, assessment of translation quality, discrimination of language complexity, etc [26].

3) Analyzing the inter-relations among different scales of human languages: the language networks can be studied from three different scale, namely the global scale, local scale and individual scale. Most researches on language networks are based on the global network and focus on the macroscopic global features of language networks. While at the local and individual scales, there are few studies. Therefore, it is an important scientific question to establish the relationship between the macroscopic overall complexity of language networks and the microscopic individual grammatical features. Some recent works have begün to address these issues. For example, Cancho et al. applied spectral analysis to study the community structure of syntactical-dependent networks and found that similar type of words have obvious aggregation behavior [27]. Siew et al. detected the community structure of the speech network, and found that the size of the community has the strong correlation with the length and frequency of words [28].

It is widely recognized that learning Chinese is much more difficult than learning Indo-European languages, especially for people whose native language belongs to the Indo-European language family. While the main difficulty lies in learning to read and write Chinese characters, that is, two most important ingredients–the root and phoneme of Chinese characters. In this work, we dedicated to improve the learning efficiency of Chinese characters via the network approach, i.e., by employing the key nodes mining algorithms, since they are rather mature and commonly applied in various situations. Rather than the traditional frequency dictionary learning strategy, we would like to provide the more complete and reasonable learning lists of roots and phonemes, and explore the interrelationship between roots and phonemes.

This paper is organized as follows. In the next section, the description of the Chinese writing systems and some linguistics concepts are introduced. In section 3, the construction of root network (RN) based on the commonly used Chinese characters is discussed, then some topological properties are presented, such as degree distribution, the hierarchical structure, the community structure, etc. The key nodes mining results are based on various quantities, such as degree, weight, centrality, percentage of network penetration, and so on. Introductions of the Chinese Pinyin and phonemes are given in section 4. In section 5, corresponding results are discussed and presented for the phoneme network (PN) of modern Chinese characters. In section 6, in order to uncover the interrelations between root and phoneme when people are learning the Chinese characters, the multi-layer network is proposed between roots and phonemes, and some learning strategies are suggested.

2. Chinese Writing System

The writing system [7] is the process or result of recording spoken language using a system of visual marks on a surface. There are mainly two types of writing systems: logographic (Sumerian cuneiforms, Egyptian hieroglyphs, Chinese characters) and phonographic. The latter includes syllabic writing (e.g. Japanese hiragana) and alphabetic writing (English, Russian, German), while the former encodes syllables and phonemes.

The unit of Chinese writing system is the character: a spatially marked pattern of strokes phonologically realized as a single syllable. Generally, each character denotes a morpheme or several different morphemes. The Chinese writing system evolved by emphasizing the concept of the character-morpheme, to some extent blurring the concept of the multi-syllable word.

Chinese characters are logograghs primarily used in the writing of Chinese, they represent words of the language using several strategies. A few characters, including some of the most commonly used, were originally pictographs, which depicted the objects denoted, or ideograms, in which meaning was expressed iconically. The vast majority were written using the rebus principle, in which a character for a similarly sounding word was either simply borrowed or (more commonly) extended with a disambiguating semantic marker to form a phono-semantic compound character.

Psycholinguistic research [6, 7] shows that the characters are important cognitive and perceptual units for Chinese writers and readers, e.g. Chinese characters are more directly related to their meanings than English words to their meanings. The explanation of this effect would be that characters (compared to English words) are perceived holistically as a meaning-carrying objects, while English words are yet to be reconstructed from a sequence of their constituents (phonemes and syllables).

3. Root Network of Modern Chinese Characters

The RN is constructed based on the commonly used 7000 modern Chinese characters from the frequency dictionary. Each character is separated into its basic root (or morpheme) according to the roots-combining-rules. For example, the Chinese character "机" can be separated as "木" and "几". We need to stress here that our current data of roots of modern Chinese characters is achieved by the authors themselves, not by the machine, therefore it is rather accurate, reliable and complete. Then we denote each root as a single node, if two or more roots co-occurrent in the same character, we assign the full links among them. By doing this for all the 7000 modern Chinese characters, the RN is constructed, with 461 nodes and 8299 links.

The degree of a node is the number of nodes connecting with this node. The degree distribution of RN is shown in Fig. 1, with $P(k) \sim k^{-1.89}$, which means the degree distribution of RN is power law distribution and the RN obeys the scale-free network. The clustering coefficient is calculated, which could represent the closeness of nodes. We study the relation between clustering coefficient and degree of RN and find that they have power law relation, $C(k) \sim k^{-0.34}$, which is shown in Fig. 2. This means that RN has the underline hierarchical structure.

Then the average shortest path distance and average clustering coefficient of RN was compared with those of the Erdős–Rényi network. Results were shown in Table 1. σ, which can be used to judge whether a network has the small-world property [29], can be calculated as $\sigma = (\frac{C}{C_r})/(\frac{L}{L_r})$, where C is average clustering coefficient of RN, C_r is average clustering coefficient of ER network, L is the average shortest path distance of RN, and the L_r is the

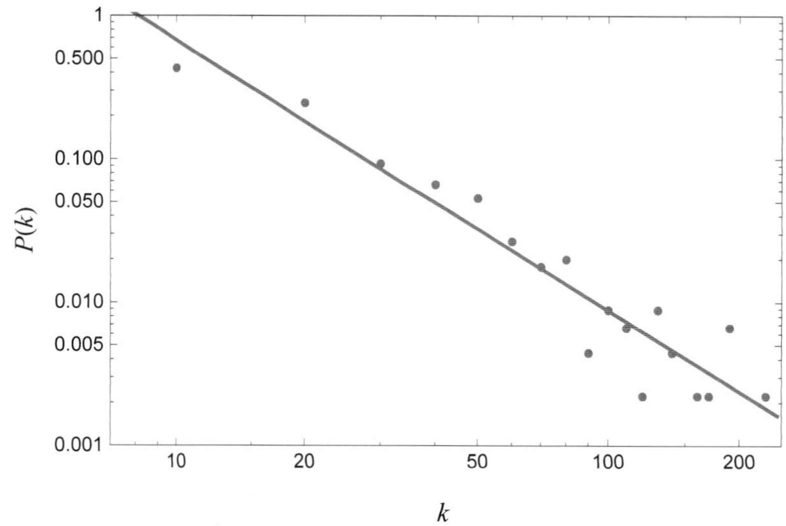

Figure 1. (Color online) The degree distribution of RN, with $P(k) \sim k^{-1.89}$. It belongs to the scale free network.

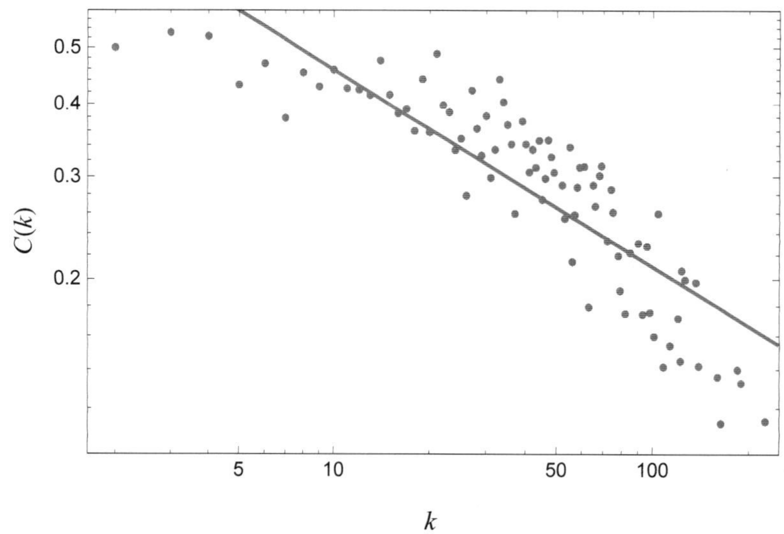

Figure 2. (Color online) The relations between the clustering coefficient and degree, $C(k) \sim k^{-0.34}$. This means that the RN has the underline hierarchical structure.

Table 1. The average shortest path distance and average clustering coefficient of RN and ER network.

Network	Average path distance	Average clustering coefficient
RN	2.24	0.39
ER network	6.18	0.052

average shortest path distance of ER network. We can get the $\sigma = 20.52 > 1$, this shows that the RN has the small-world property.

Then the centrality of RN was studied, such as the degree centrality $k_i = \sum_{j=1}^{N} a_{ij}$, eigenvector

IOP Conf. Series: Journal of Physics: Conf. Series **1113** (2018) 012019 doi:10.1088/1742-6596/1113/1/012019

Table 2. The top 15 Degree Centrality, Eigenvector Centrality, Node Weight, and Root Frequency of RN. Full list of rankings would be found via this link [39].

Character	Degree	Character	Eigenvector	Character	Weight	Character	Frequency
口	204	口	1	口	0.018	口	0.056
氵	167	木	0.926	氵	0.016	一	0.042
扌	153	艹	0.914	艹	0.015	日	0.026
木	148	氵	0.908	木	0.015	人	0.024
艹	130	扌	0.824	扌	0.014	亻	0.023
亻	114	亻	0.79	亻	0.013	白	0.022
日	110	土	0.786	钅	0.012	土	0.02
月	100	月	0.782	月	0.012	勺	0.02
土	99	日	0.747	土	0.012	木	0.019
讠	92	虫	0.743	虫	0.011	月	0.016
一	92	钅	0.74	女	0.011	扌	0.015
女	90	女	0.698	讠	0.01	辶	0.014
人	79	一	0.693	日	0.01	丷	0.012
宀	76	讠	0.663	一	0.009	寸	0.012
纟	76	人	0.659	纟	0.009	又	0.012

centrality $x_i = c\sum_{j=1}^{N} a_{ij}x_j$. The results were shown in Table 2. The average degree centrality and average eigenvector centrality of RN are 18.78 and 0.166 respectively. We also compare these results with the node weight of RN, where node weight is defined as the total weight of the edges connected with this node. One would find that the rankings of those centrality are quite similar, and the top 200 nodes occupy 81.38% weight of the network. We also compare them with the statistic result of frequency of roots in the frequency dictionary. The top 15 most frequent roots are also similar with the above results. While the top 200 most frequent roots occupy 92.1% of the network. This means that we just need to learn the top 200 roots in most cases, then going to the characters the next step.

The community structure shows the different functioning parts of the network, and the nodes in same community connect more densely. This concept was proposed by Newman [30], and the corresponding algorithm was also proposed. The popular algorithms are infomap [31], modularity maximization algorithm [32], and the Label Propagation Algorithm with compression of Flow (LPAF) [33] nowadays. The modularity maximization algorithm was chosen to explore RN, since it has the highest modularity, which could be used to judge the quality of the community structure detection. The community detection result is shown in Fig 3. One could find that the RN has very clear community structures, while the core nodes in each community are also the important nodes in the network. One would pay due attention to these nodes, and learn the roots in the communities one by one, so as to get a more systematic studying. Full lists of roots in each community would be also found via this link [39].

4. Chinese Pinyin versus Phonemes

Language can be viewed as a hierarchic construction: phoneme, syllable, morpheme, word... Each of these objects expresses meaning or participates in its formation, and consists of elements of the previous level, i.e. syllable consists of phonemes. The lowest hierarchic level is phoneme [8], which is defined to be a representative for a group of sounds that are not distinguishable with respect to their meaning-formation function in a concrete language.

The meaning is crucial for the definition of the phoneme, although a single phoneme does not express a separate meaning. The next hierarchic level (syllable) indirectly participates in the

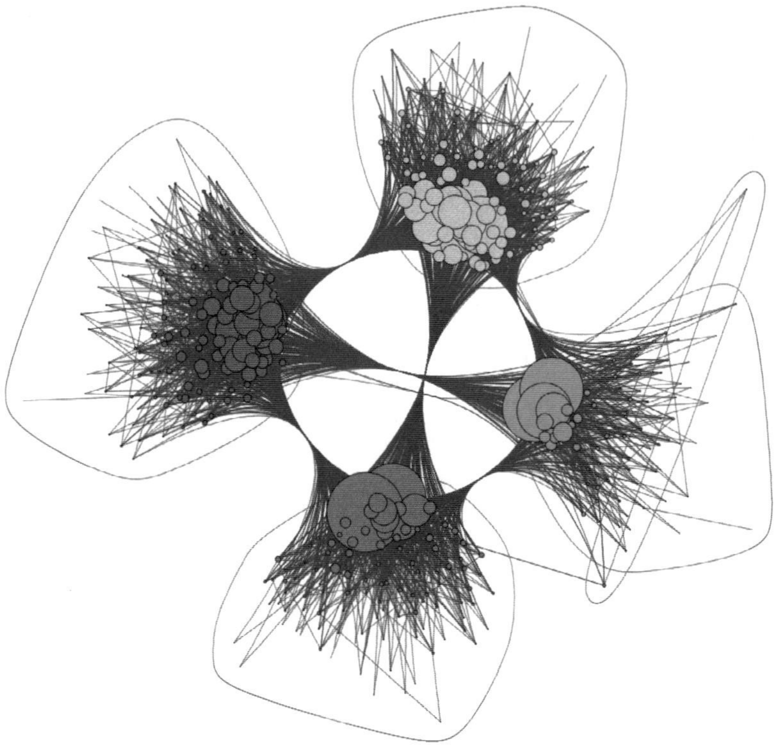

Figure 3. (Color online) The community structure of RN, nodes with the same color belong to one community, and the size of nodes implies the degree of the node.

definition of the phoneme, since the syllable bounds phonemes, i.e. there cannot be a phoneme which belongs to two different syllables; e.g. diphthongs belong to the same syllable.

Chinese language is a tonal language. The pronunciation and spelling of Chinese characters are generally given in terms of initials and finals, which represent the segmental phonemic portion of the language, rather than letter by letter. Initials are initial consonants, while finals are all possible combinations of medials (semivowels coming before the vowel), the nucleus vowel, and coda (final vowel or consonant).

5. Phoneme Network of Modern Chinese Characters

The major processes of constructing the PN are: 1) Translate the Chinese character into its corresponding Pinyin (with totally 1214 different pronunciations); 2) Split the initial consonant and simple or compound vowel as different phonemes (nodes); 3) Assign the full links among the phonemes if they co-occurrent in the Pinyin of a character. There are 161 nodes and 1350 edges in the PN.

We deal with the PN in the same way as we did with RN. The results of top 15 degree, eigenvector centrality, and node weight are shown in Table 3. We also study the community structure of PN, as is shown in Fig 4. Here we also dedicated to create the frequency dictionary of pronunciations base on the 7000 modern Chinese character (see Table 3). These results (both the key nodes mining and frequency dictionary of pronunciations) are very useful for studying the pronunciations of Chinese characters.

IOP Conf. Series: Journal of Physics: Conf. Series **1113** (2018) 012019 doi:10.1088/1742-6596/1113/1/012019

Table 3. The top 15 Degree centrality, Eigenvector centrality, Nodes weight, and Frequency dictionary of PN. Full list of rankings would be found via this link [39].

Nodes	Degree	Nodes	Eigenvector	Nodes	Weight	Pronunciation	Frequency
y	49	t	1	l	74	de	9080142
l	48	ch	0.986	t	65	o4	7034705
x	46	l	0.976	ch	63	an4	6337512
zh	44	d	0.965	d	60	e4	6297696
j	43	zh	0.95	h	59	da	5004554
d	41	sh	0.946	y	57	a1	4747652
h	41	h	0.934	m	57	e2	4490919
q	37	y	0.901	g	57	ao4	4468126
sh	37	g	0.9	b	57	shi4	4308697
t	34	n	0.821	zh	57	ai4	4140804
g	33	c	0.818	p	56	zhe	4065755
m	33	b	0.816	n	55	an1	3942001
ch	32	z	0.812	sh	55	ang4	3459766

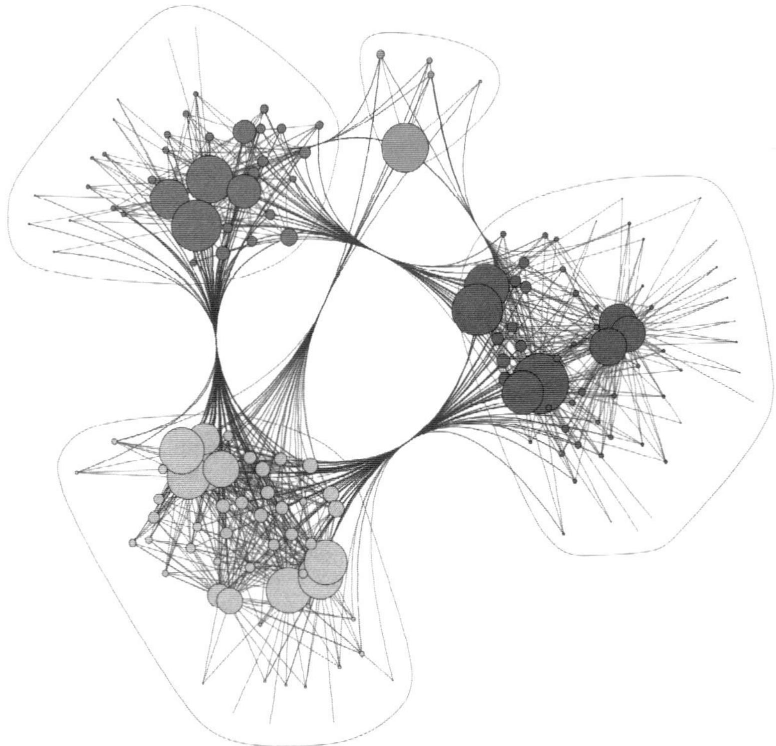

Figure 4. (Color online) The community structure of PN, in which the different colors represent different communities and the nodes size represents the degree of the node.

6. Multilayer Network of Roots and Phonemes of Modern Chinese Characters

Multilayer network, a new method for studying the relations of complex systems, has attracted more and more attention [34]. There are some common types of multilayer networks, such as interdependent networks [35], in which the characteristics of nodes in two or more monoplex networks are adjacent to each other via edges that are called dependency edges; networks of networks [36, 37], in which characteristics of each node are represented in different networks; Multiplex networks [38], a multilayer network whose layers are a sequence of graphs, different layers contain the same nodes.

We need to emphasize that roots and phonemes are not independent ingredients of Chinese characters. Therefore, the interdependent networks was employed and constructed for studying the relations of roots and phonemes of modern Chinese characters. The specific method is to set the roots and phonemes as two layers, the interconnection represents that the roots have that typical pronunciations, and the layer is same as the RN and PN. Several examples of the interdependent networks were shown in Fig. 5. It is indeed found that several roots preferentially connects to certain phonemes. Thus, this provide a potential way to study the roots and phonemes more efficiently and systematically.

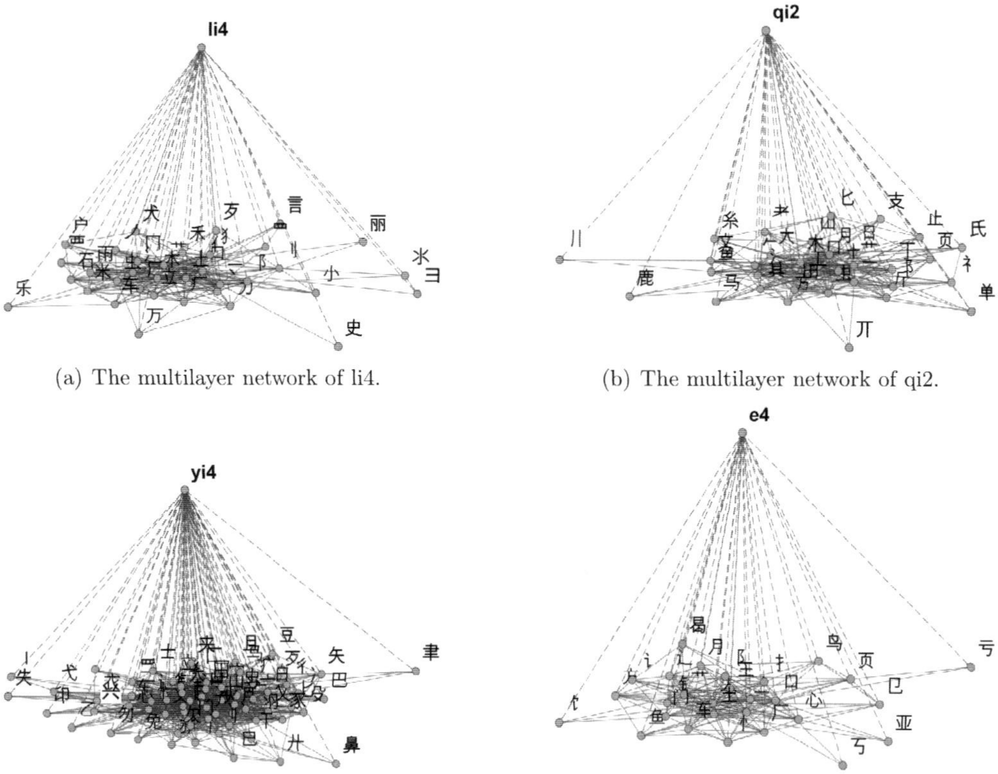

(a) The multilayer network of li4.

(b) The multilayer network of qi2.

(c) The multilayer network of yi4.

(d) The multilayer network of e4.

Figure 5. (Color online) The four examples of the interdependent networks of roots and phonemes of modern Chinese characters.

7. Conclusions

The modern Chinese characters are studied by the network approach via two basic linguistics units: roots and phonemes, in the interest of improving the learning strategy of Chinese language.

Key nodes mining algorithms are employed to dig out the important roots and phonemes, such as degree centrality, eigenvector centrality, weight and percentage of network occupation, etc. The full list of frequency dictionary of pronunciations of Chinese characters is also obtained. It is also the first time that tries to draw the relations between roots and phonemes.

References

[1] Altmann, E. G., and Gerlach, M. (2016). Statistical Laws in Linguistics. Creativity and Universality in Language. Springer International Publishing.
[2] Cong, J., and Liu, H. (2014). Approaching human language with complex networks. Physics of Life Reviews, 11(4), 598.
[3] Castellano, C., Fortunato, S., and Loreto, V. (2009). Statistical physics of social dynamics. Review of Modern Physics, 81(2), 591-646.
[4] Hu C. K., Kuo W. C. (2005). Universality and Scaling in the Statistical Data of Literary Works (POLA Forever), pp. 115–139.
[5] Chi T. T., Wang S. C., Hu C. K. (2015). Word population analysis and other evidences indicate that Shiji was amended by Liu Xiang. Physica A, 437, 408-417.
[6] Allahverdyan, A. E., Deng, W., and Wang, Q. A. (2013). Explaining zipf's law via a mental lexicon. Physical Review E Statistical Nonlinear & Soft Matter Physics, 88(6), 062804-062804.
[7] Deng, W., Allahverdyan, A. E., Li, B., and Wang, Q. A. (2014). Rank-frequency relation for chinese characters. European Physical Journal B, 87(2), 1-20.
[8] Deng, W., Allahverdyan, A. E., (2016). Stochastic Model for Phonemes Uncovers an Author-Dependency of Their Usage. PLoS ONE, 11(4) e0152561.
[9] Lecun, Yann, Y. Bengio, and G. Hinton. "Deep learning." Nature 521.7553(2015):436.
[10] Stephens, G. J., and Bialek, W. (2010). Statistical mechanics of letters in words. Physical Review E Statistical Nonlinear & Soft Matter Physics, 81(2), 066119.
[11] Bakalis, E., and Galani, A. (2012). Modeling language evolution: aromanian, an endangered language in greece. Physica A Statistical Mechanics & Its Applications, 391(20), 4963-4969.
[12] Solé, R. V. (2003). Least effort and the origins of scaling in human language. Proceedings of the National Academy of Sciences of the United States of America, 100(3), 788-91.
[13] Cancho, R. F. I. (2005). Zipf's law from a communicative phase transition. The European Physical Journal B - Condensed Matter and Complex Systems, 47(3), 449-457.
[14] Amancio, D. R., Nunes, M. G. V., Jr, O. N. O., Pardo, T. A. S., Antiqueira, L., and Costa, L. D. F. (2011). Using metrics from complex networks to evaluate machine translation. Physica A: Statistical Mechanics & Its Applications, 390(1), 131-142.
[15] Rêgo, H. H., Braunstein, L. A., D'Agostino, G., Stanley, H. E., and Miyazima, S. (2014). When a text is translated does the complexity of its vocabulary change? translations and target readerships. Plos One, 9(10), e110213.
[16] Serra J., Kantz H., Serra X. and Andrzejak R. G., Predictability of music descriptor time series and its application to cover song detection. IEEE Transactions on Audio, Speech, and Language Processing 20, 2 (2011).
[17] Basile, C., Benedetto, D., Caglioti, E., and Esposti, M. D. (2008). An example of mathematical authorship attribution. Journal of Mathematical Physics, 49(12), 237.
[18] Eduardo G A, Martin G. Physicists' papers on natural language from a complex systems viewpoint. http://www.maths.usyd.edu.au/u/ega/physicist-language/1989-2018.
[19] Liu, H. T., and Jin, C. (2013). Language clustering with word co-occurrence networks based on parallel texts. Chinese Science Bulletin, 58(10), 1139-1144.
[20] Liu, Z., Zheng, Y., and Sun, M. (2008). Complex network properties of chinese syntactic dependency network. Complex Systems & Complexity Science.
[21] Liang, W., Shi, Y., and Huang, Q. (2014). Modeling the chinese language as an evolving network. Physica A Statistical Mechanics & Its Applications, 393(1), 268-276.
[22] Cancho, R. F. I., Solé, R. V., and Köhler, R. (2004). Patterns in syntactic dependency networks. Phys Rev E Stat Nonlin Soft Matter Phys, 69(1), 051915.
[23] Motter, A. E., de Moura, A. P., Lai, Y. C., and Dasgupta, P. (2002). Topology of the conceptual network of language. Physical Review E Statistical Nonlinear & Soft Matter Physics, 65(2), 065102.
[24] Lieberman, E., Michel, J. B., Jackson, J., Tang, T., and Nowak, M. A. (2007). Quantifying the evolutionary dynamics of language. Nature, 449(7163), 713-6.
[25] HaiTao Liu, and WenWen. (2010). Language clusters based on linguistic complex networks. Chinese Science Bulletin, 55(30), 3458-3465.

[26] Amancio, D. R., Aluisio, S. M., Jr, O. N. O., and Costa, L. D. F. (2013). Complex networks analysis of language complexity. Epl, 100(100), 58002.
[27] Ramon Ferrer i Cancho, Andrea Capocci, and Guido Caldarelli. (2007). Spectral methods cluster words of the same class in a syntactic dependency network. International Journal of Bifurcation & Chaos, 17(07), 2453-2463.
[28] Siew, C. S. (2013). Community structure in the phonological network. Frontiers in Psychology, 4(4), 553.
[29] Humphries M D, Gurney K. Network 'small-world-ness': a quantitative method for determining canonical network equivalence. Plos One, 2008, 3(4): e0002051.
[30] Girvan M, Newman M E J. Community structure in social and biological networks. Proc. Nat. Acad. Sci., 2002, 99(12):7821-7826.
[31] Rosvall M, Bergstrom C T. Maps of Random Walks on Complex Networks Reveal Community Structure. Proc. Nat. Acad. Sci., 2007, 105(4): 1118-1123.
[32] Newman M E. Fast algorithm for detecting community structure in networks. Phys. Rev. E, 2004, 69: 066133.
[33] Han J, Li W, Su Z, et al. Community detection by label propagation with compression of flow. European Physical Journal B, 2016, 89(12):272.
[34] Domenico M D, Biamonte J. Spectral entropies as information-theoretic tools for complex network comparison. Physical Review X, 2016, 6(4).
[35] Kivelä M, Arenas A, Barthelemy M, et al. Multilayer networks. Ssrn Electronic Journal, 2013, 2(3): 261-268.
[36] Gao J, Buldyrev S V, Stanley H E, et al. Networks formed from interdependent networks. Nature Physics, 2012, 8(1): 40-48.
[37] Ferreira, M. A. M. (2014). Networks of Networks: The Last Frontier of Complexity-A Book Review. International Journal of Latest Trends in Finance and Economic Sciences, 4(1), 708.
[38] Battiston, F., Nicosia, V., & Latora, V. (2013). Metrics for the analysis of multiplex networks. Physical Review E, 89, 032804.
[39] https://github.com/cwj31415926/Chinese-language-data

Uncovering the role of signs' configuration in epidemic spreading on a signed social network

Fujuan Gao, Long Guo, Zhongjie Luo

School of mathematics and Physics, China University of Geosciences (Wuhan), Lumo Road, 430074, Wuhan, China

E-mail: lguo314@163.com (Long Guo)

Abstract. In this paper, we introduce a reshuffling approach to analyze how signs' configuration affects the SIR disease-spreading process on the voting network of Wikipeida (VNW) including both positive and negative edges. Inspired by the ensemble theory in statistical mechanics, many reshuffled VNWs with different signs' configuration are built through applying our reshuffling approach with a pair of parameters (r_s, r_+) on the real VNW. Through comparing SIR simulations on those reshuffled VNWs, we find that the real signs' configuration slows down disease spreading. Interestingly, there exists the suppressed effect of the fraction of recovered nodes ρ_R in the real VNW, which is a clear fingerprint of non-random factor that prevents the spreading of disease from infectious individuals to susceptible ones in human behaviors. Our present work provides a valuable perspective and tool to understand the tightly coupling of dynamical process and topology in complex system.

1. Introduction

Recently, complex networks have undergone a remarkable development and have emerged as an invaluable tool to describe and quantify complex systems in physics, biology and sociology. Generally, complex network is always assumed to have only positive interactions (i.e., friend, collaboration or trust) when we analyze the dynamics of and on complex network. However, many real complex networks[1, 2], especially online social networks (such as the Ebay, Slashdot and Wikipedia), also have negative links as well as positive ones, such as distrust, disproval and foe such as like and dislike in Slashdot, trust and distrust in Epinions and support and oppose in the voting network of Wikipedia(VNW). Social networks with both types of interactions are called signed networks, where edges are labeled with positive and negative signs[3].

The two fundamental issues are how signs are organized in real signed social network and how signs' configuration dose affect the dynamics of and on signed social networks. For the first issue, there are two different theories, one is the social balance theory[4, 5, 6, 7, 8], and the other is the status theory[1, 2]. The social balance theory, which follows the common principles that "the friend of my friend is also my friend", "the enemy of my friend is also my enemy", "the friend of my enemy is my enemy" and "the enemy of my enemy is my friend", was formulated by Heider to understand the signs' configuration in undirected signed social network from the aspect of social psychology. Facchetti et al. analyzed the global level of balance in online signed social networks, such as Epinions and Slashdot, and found that those undirected signed networks are indeed extremely balanced[9]. While for directed signed social networks, the status theory, developed by Leskovec and his collaborators[1, 2], explained signs' configuration. According to the status theory, a positive edge (u,v) means that u regards v as having higher status than himself/herself while a negative edge (u,v) means that u regards v as having lower status than himself/herself in a directed signed social network.

Commonly, individuals' interactions and states are the two basic components in society. Different types of interactions show different roles in contagion processes (such as spreading of diseases, rumors or opinions), and the distribution of types of individuals' interactions in signed social network provides a set of potential spreading channels to guide a given contagion process. While, types of individuals' interactions will be affected by the corresponding individuals' states. In detail, the evolution of edge signs can be affected by the evolution of nodes in signed social networks. Recently, some dynamical models have been proposed to analyzed signs' evolution and contagion processes on signed social networks[10, 11, 12, 13, 14, 15, 16, 17, 18]. For example, Facchetti and his collaborators analyzed signs' organization by exploring the landscape of near-optimal structure balance[10]. Fan et al. analyzed the opinion spreading based on the SIR model in homogeneous signed networks[14]. Altafini studied the process of opinion forming in structurally balanced signed networks[19]. Saeedian et al. introduced an energy function to describe the coupling problem of the sign evolution and the SI disease-spreading process based on the social balance theory and they found that balanced clusters act as obstacles to contagion processes[18]. Torok et al. divesed an extend opinion dynamic model to describe the rise, persistence and solution of conflicts on Wikipedia and found a transition from mainly consensus to a perpetual conflic occur when agents are replaced with new ones at a certain rate[20]. Y.zha et al. found a universal double-power-law distribution of time intervals between consecutive updates of an article by a set of empirical analysis of the history of millions of updates in Wikipedia and then propose a generic model to unfold collaborative human activities[21]. However, the study of signs' evolution and disease spreading in realistic signed social networks, such as Epinions, Slashdot and Wikipedia, also remains a big challenge, since those online social networks provide a paramount platform to shape agents' behavior in society.

In this paper, we focus on the issue that how signs' configuration in the VNW affects a disease-spreading process, which is a common phenomenon of diffusion process in society[22]. We consider the Susceptible-Infected-Recovered (SIR) model on a given signed social network with both positive and negative edges. In the standard SIR model, one of the most investigated and classical disease-spreading models, individuals can be in one of three states depending on the stage of the disease: susceptible (S), infectious (I) and recovered (R). During the epidemic spreading, susceptible node can be infected by his/her infectious neighbors with a transmission rate λ in one contact with any infected neighbors, infectious node can recover from the disease and then become immune[23, 24, 25, 26] with rate μ. Considering difference from a simple complex network with only positive edges, signed social network has two opposite types of edges that play different roles in affecting the SIR disease-spreading process. In this context negative edges play a block role in individuals' interactions and we assume that only positive edges have the ability to transmit disease and negative edges do not transmit disease[18]. Hence, signs' configuration, i.e., a given distribution of positive and negative signs, provides a set of potential spreading channels and affects the SIR spreading process. The main question is how to rebuild different signs' configurations as samples. Inspired by the ensemble theory in statistical mechanics, we propose a reshuffling approach that operates on edges of the VNW,then obtain different reshuffled VNWs related to parameters r_s and r_+. Through comparing SIR simulations on different reshuffled VNWs, we find that the real signs' configuration prevents the spreading of infection over the whole network and there exists an obvious suppress effect to describe the positive role of non-randomness in the real signs' configuration.

The paper is organized as follows. In Section 2, signed social networks and our reshuffling approach are defined, and then we define a SIR disease-spreading process on different reshuffled signed networks. In Section 3, we compare SIR simulations on different VNWs and get our

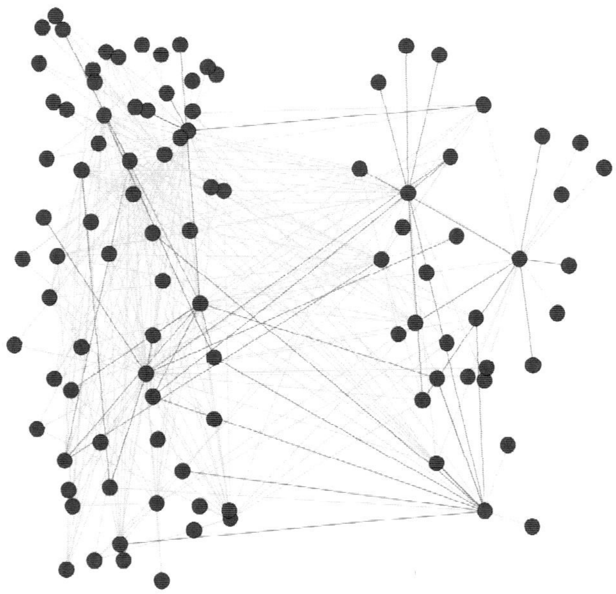

Figure 1. (color online) The illustration of a subgraph of the real VNW. Red and green links are the links with the negative and positive sign respectively.

interesting results. Finally, the conclusion and discussion are given in Section 4.

2. The SIR model on reshuffled signed networks

To begin with, we consider a directed signed social network $G = (V, L, A)$, where V is the set of vertices, L is the set of directed links, $A = \{A_{uv}\}$ describes the signed adjacency matrix with $A_{uv} \neq 0$ if and only if $(u, v) \in L$ and A_{uv} quantifies the sign of edge (u, v). A positive sign $A_{uv}(= +1)$ represents that u tags v as a friend or u supports v, while a negative sign $A_{uv}(= -1)$ represents that u tags v as enemy or u opposes v. Here, we focus on analyzing the role of signs' configuration of the voting network of Wikipedia (VNW) in disease-spreading. The VNW is a network of users that voted for and against each other in admin elections, including $N_v = 7118$ users and $N_l = 103,747$ edges where the proportion of positive edges is $r_+^0 = 78\%$[1, 2]. So the VNW network can be described to signed network. We use the data of the voting network of Wikipeida (VNW) to establish signed network and simulate SIR disease-spreading process, which 7118 users on the VNW network are mapped to 7118 nodes in signed network and relationships between users are abstracted to 103,747 edges including 78% positive edges. In Fig.1, a subgraph of the real VNW is shown as a undirected signed network, where red and green edges represent the edges with negative sign and positive sign, respectively.

In order to analyze signs' configuration in the real VWN from a viewpoint of dynamics, we choose the SIR model as a disease spreading process on the VNW. Each node, says i, is labeled by S_i as its state, where $S_i = 0$ if the node is susceptible, $S_i = 1$ if it is infected and $S_i = -1$ if it is recovered and gets immunization. As mentioned above, the two opposite types of edges play different roles during a given scenario, and we assume that positive edges can transmit disease and negative edges cannot. Hence, a SIR disease-spreading process depends not only on nodes' states but also on edges' signs, since different signs arrangement provides different disease-spreading channels.

We then propose a reshuffling approach that operates on edges and rebuild the reshuffled VNWs with different signs' configuration according to two tuning parameters (r_s, r_+). Our reshuffling approach is introduced based on the ensemble theory in statistical mechanics, since a given signed social network with constant macroscopic quantities (N_v, N_l) can be regarded as an isolated system obeying the ergodic hypothesis. A signed social network can have many microscopic states according to signs' configuration. To reset one or more signs will alter $itssigns'$ **sign's** configuration, and the system will experience another microscopic state. Different signs configuration describes different microscopic state under the same topology structure of a given system. It's sure there exist one specific microscopic state which has the same signs configuration as the real signed social network.

In detail, our reshuffling approach is described that each edge is chosen as an activity one with the reshuffling probability r_s, and then each activity edge resets its sign as positive with probability r_+ and as negative with probability $(1 - r_+)$. We can obtain 2^{N_l} potential reshuffled signed networks with different microscopic states according to the tuning parameters (r_s, r_+). Each reshuffled signed network provides its own set of disease-spreading channels. Obviously, the fraction of positive signs of a reshuffling VNW with parameters (r_s, r_+) can be written as

$$R_+ = r_s r_+ + (1 - r_s) r_+^0 \tag{1}$$

Specially, a reshuffled signed network reduces to the real one when $r_s = 0$, while all edges' signs are reshuffled randomly when $r_s = 1$. Note that our reshuffling approach provides the possibility and feasibility to compare roles of different signs' configurations in contagion processes, especially the SIR disease-spreading process in this paper.

For a given reshuffled VNW with parameters (r_s, r_+), we start the SIR disease-spreading process from an initial state, where a ratio of notes which were randomly infected is ρ_0 and others are susceptible. At each time step, each susceptible node is infected with rate λ if it is connected to one or more his/her infected neighbors along the positive edges. At the same time, infected nodes are cured and acquire immunity with rate μ. Without lack of generality, we can set $\mu = 1$. After enough time elapses, there does not exist any one infected nodes in a given system and we are interested in the fraction ρ_R of recovered nodes when the SIR disease-spreading process is finished. In order to reduce fluctuation, the SIR disease-spreading process is repeated $N_c = 50$ times independently with the same initial condition. And we record the evolution of $\langle \rho_R \rangle (= \frac{1}{N_c} \sum \rho_R)$ as a function of the infected rate λ. Besides, we will get many reshuffled VNWs by means of our reshuffling approach with tuning parameters (r_s, r_+), and repeat the SIR disease-spreading process on those reshuffled VNWs. We shed light on the evolution of $\langle \rho_R \rangle$ as functions of r_s and r_+ respectively.

3. Results

The primary interest is how signs configuration does affect the epidemic spreading on the VNW with its topology structure fixed. As show in Fig.2, the fraction of recovered nodes $\langle \rho_R \rangle$, an order parameter, evolves as a function of infection rate λ for different resetting probabilities r_+ and $r_s = 0.5$. Clearly, $\langle \rho_R \rangle$ increases as λ increases, which means that the larger the infection rate is, the wider the epidemic spreading is. Also, as expected, for a case of given λ, we find that ρ_R also evolves as an increasing function of r_+, which indicates that more potential spreading channel promotes the epidemic spreading. In order to quantify the difference between signs' configurations, we assume that similar signs' configurations, i.e, the similar potential spreading channels, have the same role in affecting the contagion process of the SIR model. Hence, a quantity $\Delta_g(\lambda, r_s, r_+)$ (see Fig.2) is introduced to describe the difference of signs' configuration in a reshuffled VNW and that in the real VNW under the tuning parameters λ, r_s and r_+. By this definition, $\Delta_g(\lambda, r_s, r_+)$ is given by

$$\Delta_g(\lambda, r_s, r_+) = |\langle \rho_R \rangle (\lambda, r_s, r_+) - \langle \rho_R \rangle (\lambda, r_s = 0, r_+)| \tag{2}$$

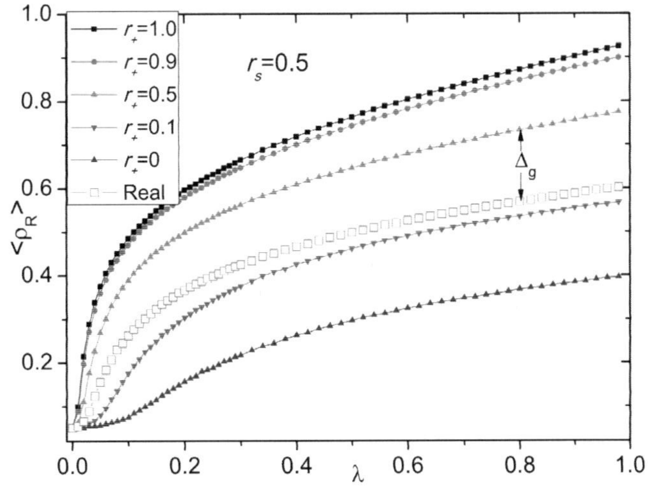

Figure 2. (color online) The ratio of recovered nodes $\langle \rho_R \rangle$ evolves as a function of the infected ratio λ under different r_+s when $r_s = 0.5$ is fixed. Δ_g is a recovered gap to quantify the difference extent of the role of the two corresponding signs' configuration in the SIR spreading dynamics for a given λ.

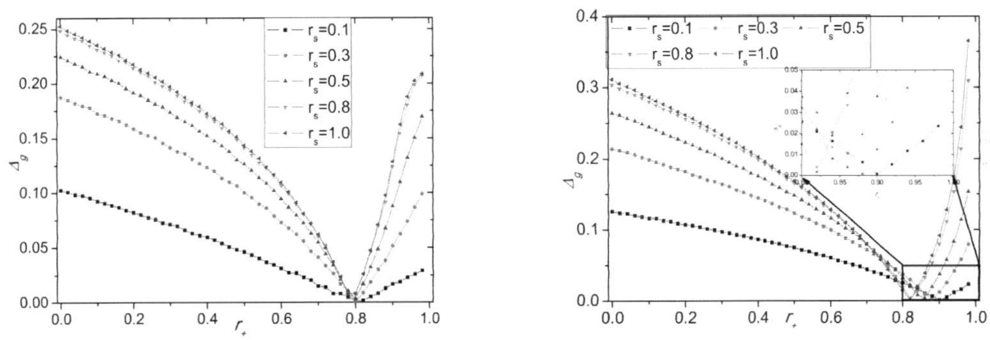

Figure 3. (color online) The relationship between Δ_g and r_+ under different r_s when the infected ration $\lambda = 0.1$ (left) and $\lambda = 0.5$ (right). The inset is to zoom in the subpart in the black rectangle in the right figure.

where, $\langle \rho_R \rangle(\lambda, r_s = 0, r_+)$ quantifies the fraction of recovered nodes in the real VNW, and the resetting probability r_+ is an invalid variable since no edge is chosen as an activity one to reset its sign when $r_s = 0$. Note that the smaller the value of $\Delta_g(\lambda, r_s, r_+)$, the more similar between the signs' configuration in the reshuffled VNW and that in the real VNW.

In Fig.3, we plot that Δ_g evolves as a function of r_+ under different r_s in the case of low infection ratio $\lambda = 0.1$ (left) and high infection ratio $\lambda = 0.5$ (right) respectively. Obviously,

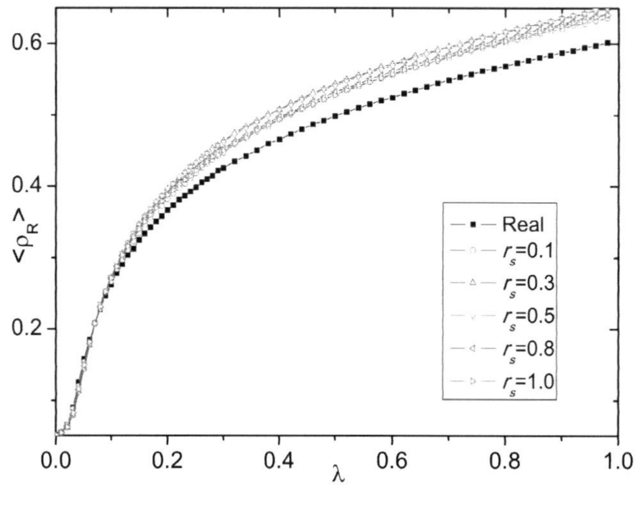

Figure 4. (color online). The fraction of removed nodes ρ_R evolves as a function of the infection ratio λ under different r_s when $r_+ = 0.78$ is fixed.

all curves have the same evolution tendency to decrease to the valley first and then increase as r_+ increases. The value of Δ_g at valley reaches zero, which means that signs' configuration in a reshuffled VNW with a given pair of parameters (r_s, r_+) has the same role as that in the real VNW in epidemic spreading. While Δ_g increases as r_s increases when r_+ is fixed, which indicates that the performance of our reshuffling approach provides more and more epidemic spreading channels with r_s increasing. Furthermore, all curves reach the same minimal value when $r_+ = 0.8$ in the case of $\lambda = 0.1$ in the left picture of Fig.3, and $r_+ = 0.8$ is approximately equal to the real fraction of positive sign 0.78 in the VNW. Surprisingly, when we increase λ, taking $\lambda = 0.5$ in the right picture of Fig.3 for example, the positions where Δ_g reaches its minimal value are separation. $r_+|_{\Delta_g \simeq 0}$ increases as r_s decreases. For $\lambda = 0.5$, Δ_g reaches its minimal value at $r_+ \simeq 0.9$ when $r_s = 0.1$. As r_+ increases, the fraction of positive edges increase, that is, potential disease spreading channels increase. At this point, the rate of recovered node in reshuffled network is the same as that of the real network. In other words, there is non-randomness that prevents the spreading of disease in the real VNW network.

Next, we focus on the issue why our reshuffling approach can open additional spreading channels. Unlike atoms based on the least action principle, individual usually maximizes his/her benefits through complex psychological activity. We assume that there exist random factor and non-random factor in individual's behavior from the macroscopic viewpoint, just like the coexistence of rational thinking and perceptual thinking in psychological activity. Then we define W_{ran} and W_{nonran} to quantify random factor and non-random factor and obtain

$$W_{ran} + W_{nonran} = 1 \tag{3}$$

Random factor is added in the reshuffled signs' configuration and the weight of random factor W_{ran} increases as r_s increases due to the definition of our reshuffling approach. All signs are reshuffled thorough randomly and $W_{ran} = 1$ when $r_s = 1$.Specially,if there is no non-random factor in real signs'configuration, i.e., $W_{ran} = 1$, at this time, all curves in Fig.4 should be

IOP Conf. Series: Journal of Physics: Conf. Series **1113** (2018) 012020 doi:10.1088/1742-6596/1113/1/012020

overlapped with each other. In order to reveal the role of non-random factor in epidemic spreading, we should suppose that the fraction of positive sign in any potential reshuffled VNW must be fixed and equal to that in the real VNW when we introduce our reshuffling approach to the real VNW. In Fig.4, we plot the evolution of $\langle \rho_R \rangle$ as a function of the infection ratio λ under different r_s when $r_+ = 0.78$. It's shown that $\langle \rho_R \rangle$ in a reshuffled VNW with $r_s \neq 0$ is larger than that in the real VNW ($r_s = 0$) when $\lambda > 0.2$, which indicates that real signs' configuration slows down the epidemic spreading.As we already mentioned that our reshuffling process is the removing process of non-random factor, the suppressed effect of $\langle \rho_R \rangle$ is an obvious fingerprint of the non-random factor in real signs' configuration according to the epidemic spreading. Furthermore, it is interesting that all curves are overlapping with each other when $r_s \neq 0$, which shows that role of non-random factor in real signs' configuration in the epidemic spreading is thoroughly released when we analyze a part of the real VNW.

4. Conclusions

In summary, we have introduced a reshuffling approach to compare SIR simulations on different reshuffled VNWs including both positive and negative edges. For simplicity, we assume that only positive edges can transmit disease when considering a SIR disease-spreading process. Signs' configuration, i.e., a distribution of positive and negative edges, provides its own set of potential disease-spreading channels to guide disease transmission. Hence, to rearrange signs' configuration is our primary task when we analyze roles of different signs' configuration in a disease-spreading process. Inspired by the ensemble theory in statistical mechanics, we propose a reshuffling approach to rebuild VNWs through tuning parameters r_s and r_+. Through analyzing the order parameter $\langle \rho_R \rangle$ in different reshuffled VNWs, we find that real signs' configuration slows down the SIR disease-spreading, since individuals have complex psychological activity, which is labeled as a non-random factor in society. Our present work provides an interesting perspective and tools to understand the coupling interaction between contagion processes and topological structure in signed social networks.

References

[1] J. Leskovec , D. Huttenlocher and J. Kleinberg 2010 *Sigchi Conference on Human Factors in Computing Systems ACM.* 1361-1370
[2] J. Leskovec , D. Huttenlocher and J. Kleinberg 2010 *International Conference on World Wide Web ACM* 641-650
[3] Y. Chen , X. L. Wang and B. Yuan and B. Z. Tang, 2014 *J. Stat. Mech.: Theo. and Exp* **03** 03021.
[4] F. Heider 1946 *The Journal of Psychology.* **21** 107-112
[5] T. Antal , P. L. Krapivsky and S. Redner 2005 *Phys. Rev. E* **72** 036121
[6] S. A. Marvel , S. H. Strogatz and J. M. Kleinberg 2009 *Phys. Rev. L* **103** 198701
[7] S. A. Marvel , J. Kleinberg , R. D. Kleinberg and S. H. Strogatz 2011 *Proc. Natl. Acad. Sci. USA* **108** 1771
[8] A. V. Rijt 2011 *Journal of Mathematical Sociology* **35** 94-113
[9] G. Facchetti , G. Iacono and C. Altafini 2011 *Proc. Natl. Acad. Sci. USA* **108** 20953-20958
[10] G. Facchetti , G. Iacono and C. Altafini 2012 *Phys. Rev. E* **86** 036116
[11] T. H. Summers and I. Shames 2013 *Europhys. Lett.* **103** 18001
[12] R. Nishi and N. Masuda, 2014 *Europhys. Lett.* **107**,48003.
[13] S. Tan and J. Lü, 2016 *Sci. Rep.* **6**, 22022.
[14] P. Fan , W. Wang , P. Li and Z. Jiang, 2012 *J. Stat. Mech. Theory Exp.* **8**,08003.
[15] S. Righi and K. Takács , 2014 *Adv. Complex Syst.* **17**, 1450011.
[16] M. Ehsani and M. M. Sepehri 2014 *Journal of Industrial and Systems Engineering* **7** 104-117
[17] J. Q. Jiang 2015 *Phys. Rev. E* **91** 062805
[18] M. Saeedian , N. Azimi-Tafreshi , G. R. Jafari and J. Kertesz 2017 *Phys. Rev. E* **95** 022314
[19] Altafini Claudio 2012 *PLoS ONE* **7** 38135
[20] Török J, Iñiguez G, Yasseri T, San, M. M., Kaski, K., and Kertész, J 2012 *Phys. Rev. L* **110** 088701
[21] Zha, Y., Zhou, T., and Zhou, C. 2016 *Proceedings of the National Academy of Sciences of the United States of America* **113** 14627.
[22] R. Pastor-Satorras , C. Castellano , P. V. Mieghem and A. Vespignani 2015 *Rev. Mod. Phys.* **87** 925-979

[23] W. O. Kermack and A. G. McKendrick 1927 *Proceeding of the Royal Society of London A* **115** 700-721
[24] J. Zhou and Z. H. Liu 2008 *Front. Phys. China* **3** 331-348
[25] A. L Lloyd and R. M. May 2001 *Science* **292** 1316-7
[26] T. Hasegawa and K. Nemoto 2016 *Phys. Rev. E* **93** 032324

Investigate the relationship between learners' social characteristics and academic achievements

Zhi Liu[1,2], Lingyun Kang[1], Zhu Su[1,2*], Sannyuya Liu [1,2], Jianwen Sun[1,2]

[1]National Engineering Research Center for E-Learning, Central China Normal University, Wuhan, P. R. China

[2]National Engineering Laboratory for Technology of Big Data Applications in Education, Central China Normal University, Wuhan, P. R. China

E-mail: suz@mail.ccnu.edu.cn

Abstract. The social interaction behaviors of group learning have obtained a lot of attention from researches, and social network analysis (SNA) plays a critical role in exploring collective interactive patterns. In this paper, we investigated the experimental data from the discussion forum of "Discourse Analysis" course in a university learning platform. By calculating learners social characteristics, we explored the relationship between social characteristics and academic achievements, as well as the differences between various achieving learners. The results show that there exist strong positive correlations between most social characteristics and academic achievements, and the majority of high-achieving learners are located at the core position of the network. In addition, the high-achieving learners exhibited more active participations in discussions and received more posts compared with middle-and low- achieving learners.

1. Introduction

As the focus on online learning continues unabated, Massive open online courses (MOOCs) like Coursera, Udacity, edX and Khan Academy provide learners with better learning support and experience to some extent [1, 2]. To adapt MOOCs to higher education, a variant of MOOCs- Small Private Online Course (SPOCs) have emerged to support the online learning of on-campus students locally [3]. Teachers often conduct interactive teaching through SPOC forums, in which learners can consititue various learning communities for collaborative learning, ideas-sharing, and help-seeking from peers [4]. There is a large amount of data generated in the process of knowledge construction, and social network analysis (SNA) is one of the most important methods in the analysis of data collected through online learning communities. By SNA method, we can understand how the learners interact with each other and what role each learner plays in the network. Besides, by uncovering the interactive factors between learners, learners' knowledge construction and the development of higher-order thinking may be effectively improved. Therefore, it is worth to investigate that, what the relationship between social characteristics and academic achievements is, and what the social characteristics differences between various achieving learners are. The exploration of these questions is beneficial for teachers timely adjust resource allocation and teaching strategies to provide better personalized services for learners. This paper aims to adopt social network analysis method to carry out the empirical research to reveal the relationship between social characteristics and academic achievements, as well as the differences of social characteristics between various achieving learners. The remainder of this paper is organized as follows: In Section 2, we review the definition of social network analysis and related research in the online learning environment. Section 3 describes the design of this study. Results are detailed in Section 4, followed by the conclusion in Section 5.

2. Related works

Social network analysis is a method and technique used to study the interactivity among members within the exchange of resources. Members are represented as the nodes in the network, and flow of information between members constitute the links between nodes [5]. SNA has been first applied to the field of mathematics, and then to other fields by sociologists and psychologists in the first half of the 20th century. It provides a quantifiable means for the behavioral characteristics of group learning [6]. A significant number of researches have been dedicated to capture more insight into organizational phenomena through SNA approach. Oleksandra & Shane integrated SNA, discourse content analysis and statistical network modelling to jointly explore the social processes among a group of learners with relatively continuous contributions to a MOOCs course [7]. Wise et al. suggested that different tie definitions and content/non-content networks both have an impact on social network structures in MOOCs discussion forums [8]. Liu et al. adopted SNA to explore how elementary learners collaborate with peers to create stories and discovered that the knowledge level of learners has a certain influence on their position in the network [9]. Laningawijnen et al. utilized SNA to verify whether peer norms for achievement goals have an impact on friendship selections or not [10]. Recently, some researches focused on exploring the relationship between social characteristics and academic achievements. For instance, Williams et al. measured student centrality during the semester to predict future academic performance [11]. Gitinabard et al. adopted learners' online interaction data establishing the directed network to investigate how the learners' social characteristics are correlated with academic achievements in three blended courses [12]. Related ideas were pursued by Lee & Bonk. [13] and Liu et al [14]. Houston et al. compared the correlations between course final grade and engagement patterns, and suggested that the number of interactions have stronger correlations with final grade [15]. Joksimovi et al. investigated how the factors generated during learners interactions within two instances on the MOOCs platforms have the influence on the relationship between the three social network centrality measures (i.e., *degree*, *closeness*, and *betweenness*) and final course outcomes [16].

3. Research Design

3.1. Forum Dataset

In this paper, we have collected the interaction data from StarC, which is a learning platform and issued by National Engineering Research Center for E-Learning of Central China Normal University (CCNU). The experimental data comes from the discussion forum within the "Discourse Analysis" course that were delivered on the StarC platform in the Spring of 2016. The course lasts for 4 months and presented by integrating the classroom teaching and online collaborative discussions. In total, there are 134 learners who are registered in this course across a semester, in which 115 learners participated in course discussions and published 1,358 posts. In addition, the academic achievement of each participant is indicated by the final score given at the end of a semester, which is composed of the usual score and the final score. The average score of these learners reaches 83.8 with the scale of $0 \sim 100$.

3.2. Research Questions

Investigating the relationship between social interactions and academic achievements contributes to understand the potential causes behind different levels of achievement among learners, thus it is necessary to analyze the interaction patterns between various achieving learners. In this experiment, the discussion posts published within the "Discourse Analysis" course forum will be used to explore the following two aspects: (1) Relationship between social characteristics and academic achievements. (2) Differences on network positions of various achieving learners, as well as the social characteristics differences among various achieving learners. In sum, this paper aiming at addressing the following questions:

(1) What is the relationship between social characteristics and academic achievements?
(2) What are the differences between network positions of various achieving learners?
(3) What are the differences of social characteristics between various achieving learners?

3.3. Instruments and Methods

3.3.1. Social characteristics The network is composed of individuals and the interactions between them. It is the exchange of resources among individuals that make the information flow from one direction to another. Therefore, it is very necessary to know how individuals in the network control the frequency and direction of communication can enable the correct dissemination of information. SNA provides a quantifiable means, which are conducive to discover what roles individuals play in the network. The quantitative indicators are explained as follows:

Table 1. Explanations about social characteristics.

Indicators	Implications
Authority	Learners with important influence in the network, it is also can be quantified by *indegree* centrality of the corresponding node.
Betweenness	Learners who on the shortest path connecting two other learners.
Closeness	The length of shortest path between the learner and others.
Clustering	It means how well the learners neighbors are connected to each other in the network.
Degree	The frequency a learner communicates with other learners. In this experiment, it is also quantified as the total number of learner posts.
Eigenvector	It reveals the importance of learners in the network, and a learner can gain indirect influence from the high-eigenvector learners around him.
Hub	Learners with key links in the network, it is also can be quantified by outdegree centrality of the corresponding node.
Indegree	Number of posts received by a learner.
Outdegree	Number of posts sent by a learner.
Pagerank	A learners weighted popularity, reputation, or influence within a group [11].

3.3.2. The division of various achieving learners Kelley proposed that the optimal choice for dividing upper and lower groups is based on 27% of the score distribution [17], thus we divide all the learners ($N = 134$) on the "Discourse Analysis "course into high- ($N = 36$, $score >= 88$), middle- ($N = 62$, $81 < score < 88$) and low-achieving ($N = 36$, $score <= 81$) groups. It is worth to note that not all learners are involved in forum discussions, Among the 115 participants, there are 35 high-achieving learners, 52 middle-achieving learners and 28 low-achieving learners, respectively.

3.3.3. Data Processing Above all, the data extracted in the "Discourse Analysis" course forum would be filtered and organized, some noisy samples have been removed (e.g., posts made by

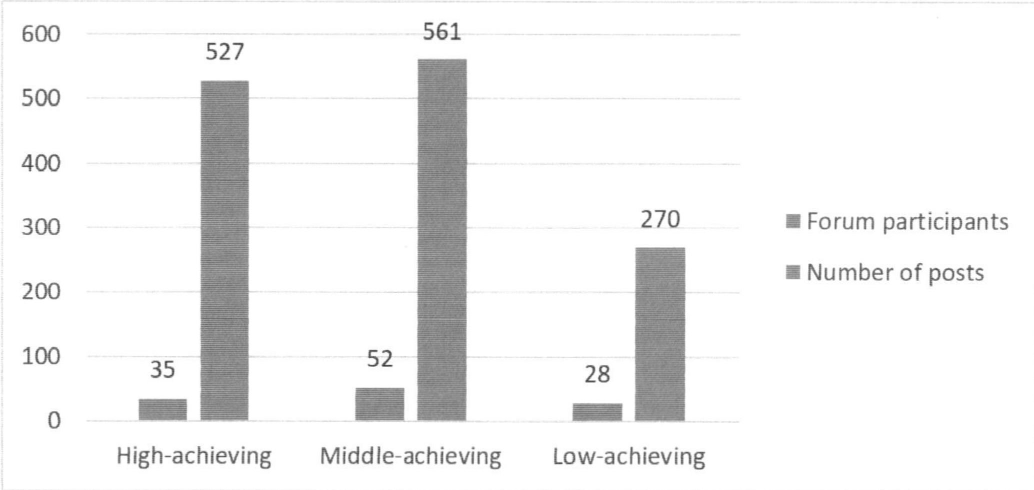

Figure 1. (Color online) Statistical results of participants and forum posts.

teachers in forums, non-registered learners posts). Then, the tool of Python's NetworkX Package has been used to calculate each participants social characteristics, and the Gephi visual software has been adopted to analyze network positions of various achieving learners. In the end, to demonstrate the relationship between social characteristics and academic achievements, the IBM SPSS software has been used to calculate Spearman correlation and Regression, at the same time, the Kruskal-Wallis H test and Pairwise Comparisons have done to compare the differences of social characteristics between various achieving learners.

4. Research results and analysis

4.1. Descriptive Statistics

It can be seen from Figure 1 that there are 115 learners involved in the discussions and 1,358 posts have been generated, in which 35 high-achieving learners, 52 middle-achieving learners and 28 low-achieving learners generated 527, 561 and 270 posts, respectively. It is worth to note that teachers posts are not included in the statistics. We can also observe that the participation rate of overall forum discussion is 85% and the number of posts per learner is 11.8, in which the different groups participation rate is 97%, 83% and 77%, as well as the number of posts per learner is 14.6, 9.05 and 7.5, respectively. The statistical results indicate that the forum discussion participation rate shows the same downward trend as the number of posts per learner, which further reveals that high-achieving learners participate more actively in forum than others.

4.2. Correlation Analysis and Regression Analysis

To address the first research question, the relationship between learners social characteristics and academic achievements is measured by Spearman Correlation in Figure 2. It shows that most of the social characteristics have significant relationship with academic achievements except for *clustering*, *hub* and *outdegree*. It can be unfolded that the correlation coefficient between the score and *authority* in this course reaches 0.368 ($p < 0.01$), indicating that there is a significantly positive correlation between the two variables. The *authority* can be quantified by *indegree* centrality of the corresponding node, in other words, the learners with high-authority often receives more posts from others. It makes perfect sense that when a learner has more adequate knowledge about the course content, his/her academic performance would be better,

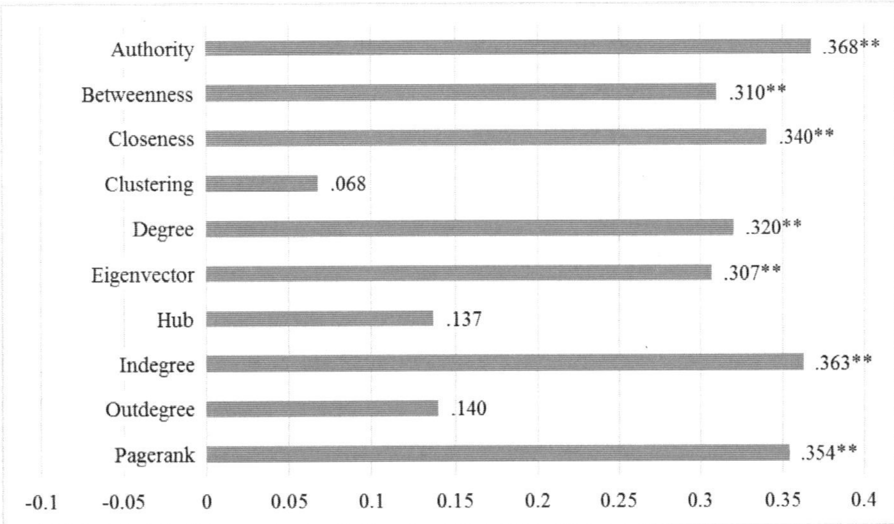

Figure 2. (Color online) Relationship between social characteristics and academic achievements.

in fact, high-achieving learners are the most popular ones to be asked for help or discussed problems by others. We also observe that there is a significant positive correlation between score and *closeness*, which means that the higher the *closeness* of a learner, the less dependence on others when the learner acquires resources. It also implies that the closer the learner is to others in the network, the better his/her academic achievements will be.

However, it can be seen from Figure 2 that *hub* and *outdegree* have weak correlations with academic achievements, which may be attributable to the inherently nature of the course. Actually, there are many differences between foreign language learning and native language learning such as words, the order of sentences, as well as cultural differences. Thus learners need to change their original linguistic thinking in order to further understand the course content, which is obviously more difficult than other non-foreign language learning courses. Consequently, the teachers of the "Discourse Analysis" program may need to provide targeted help to learners who ask questions in the forum and the learners should also ensure the sufficient time and rich content of the offline learning before sending posts online. We acknowledge that these interpretations are speculative, but the context is indeed the primary factor affecting learner behavior patterns [18], as well as the prediction of academic performance with social characteristics.

Then, the linear regression analysis was conducted to examine to what extent the social characteristics are valid indices to predict academic achievements. Among these 10 social characteristics, we only detect that eigenvector has a significant predictive relationship with academic achievements, which is shown in Table 2. Furthermore, we made a multiple regression analysis to verify whether other social characteristics can effectively predict *eigenvector* centrality or not. For this purpose, using the stepwise selection method to select more relevant factors and eliminate insignificant factors, VIF (Variance Inflation Factor) is used to verify whether there is multicollinearity between the independent variables. It can be seen from Table 3 that the variation range of the VIF value is 1.156 to 4.193 (all less than 10), which means that the multicollinearity is reasonable in this analysis. In addition, in the final linear regression equation, the independent variables explained 97.4% of the dependent variable (corrected $R^2 = 0.974$), $F = 857.532$, $p < 0.001$, which means that the fitting result is very good. The stepwise regression

Table 2. Results of linear regression analyses of predicting scores with *eigenvector* centrality.

Predictor Variables	*B*	*SE*	*β*	*t*	*Sig.*
(constant)	81.698	.947	/	86.247	.000
Eigenvector	37.376	11.488	.293	3.254	.002

Table 3. Results of multiple regression analyses for social characteristics predicting *eigenvector*.

Predictor Variables	*B*	*SE*	*β*	*t*	*VIF*
Degree	.006	.000	.844	27.331***	4.193
Outdegree	.004	.000	.225	10.316***	2.085
Clustering	.007	.002	.055	3.419**	1.156
Betweenness	-.259	.069	-.074	-3.741***	1.699
Closeness	.017	.008	.055	2.055*	3.153

$***p < 0.001$; $**p < 0.01$; $*p < 0.05$.

reduced the number of independent variables from 9 to 5, including *degree*, *outdegree*, *clustering* and *closeness* as significant positive predictors of *eigenvector*, while the *betweenness* is the significant negative predictors of *eigenvector*, and the *authority*, *indegree*, *hub* and *pagerank* were eliminated from the equation.

4.3. The Difference of Network Positions of various achieving learners

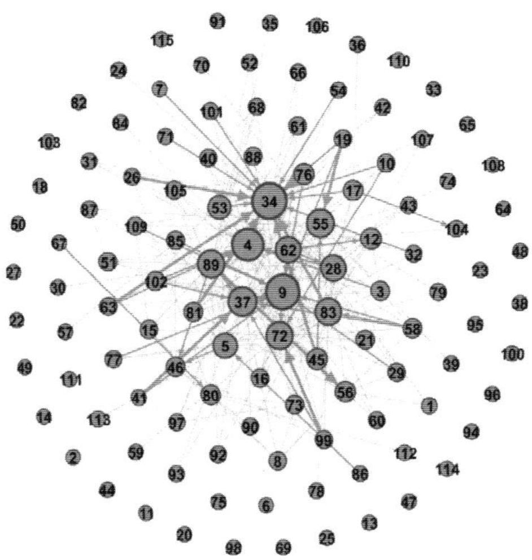

Figure 3. (Color online) Sociogram of different achieving learners.

To address the second research question, we draw the sociogram including various achieving learners distributions in social interactions within the course of "Discourse Analysis". The results are shown in Figure 3, where the orange nodes denote high-achieving learners, purple nodes denote middle-achieving learners, and green nodes denote low-achieving learners. The number on the node is marked in the order of posting. Moreover, the node size represents nodes *degree*, which is the sum of replying and delivering postings. It can be observed from Figure 3 that the majority of orange nodes are located at the central positions in the network, while

Table 4. Results of Kruskal-Wallis H test for social characteristics of high-, middle- and low-achieving learners.

Indicators	*H*	*P*
Authority	9.746**	.008
Betweenness	6.901*	.032
Closeness	8.361*	.015
Degree	10.976**	.004
Eigenvector	10.086**	.006
Indegree	9.385**	.009
Pagerank	8.826*	.012

$^{**}p < 0.01$; $^{*}p < 0.05$.

Table 5. Results of Pairwise Comparisons of high-, middle- and low-achieving learners.

Variable	*Group(I)*	*Group(J)*	*H*	*P*
Authority	Low-achieving	Middle-achieving	3.477	0.577
	Low-achieving	High-achieving	18.832**	0.005
	Middle-achieving	High-achieving	15.355**	0.008
Betweenness	Low-achieving	Middle-achieving	4.04	0.498
	Low-achieving	High-achieving	15.707*	0.015
	Middle-achieving	High-achieving	11.667*	0.036
Closeness	Low-achieving	Middle-achieving	3.404	0.585
	Low-achieving	High-achieving	17.532**	0.009
	Middle-achieving	High-achieving	14.128*	0.015
Degree	Low-achieving	Middle-achieving	9.055	0.24
	Low-achieving	High-achieving	26.504**	0.001
	Middle-achieving	High-achieving	17.449*	0.015
Eigenvector	Low-achieving	Middle-achieving	10.304	0.182
	Low-achieving	High-achieving	25.939**	0.002
	Middle-achieving	High-achieving	15.636*	0.03
Indegree	Low-achieving	Middle-achieving	3.971	0.524
	Low-achieving	High-achieving	18.743**	0.005
	Middle-achieving	High-achieving	14.772*	0.011
Pagerank	Low-achieving	Middle-achieving	3.9	0.532
	Low-achieving	High-achieving	18.204**	0.007
	Middle-achieving	High-achieving	14.304*	0.014

$**p < 0.01$; $*p < 0.05$.

most of purple nodes and green nodes are located at the edge positions in the network, which signifies that high-achieving learners are more positive in the forum. We also observe that most arrows are pointed to the center, which implies that the higher degree centrality learners often receives more posts than others.

4.4. Social Characteristics Differences between Various Achieving Learners

To address the third research question, the differences between the social characteristics of three achieving groups are examined. From Figure 2, we can observe that seven social characteristics (*authority*, *betweenness*, *closeness*, *degree*, *eigenvector*, *indegree* and *pagerank*) have significant relationship with academic achievements. Furthermore, the Kruskal-Wallis H test is utilized to explore the differences between these seven social characteristics of high-, middle- and low-achieving learners. The results are presented in Table 4, which reveals that various achieving learners significantly differ in these seven social characteristics. For further analysis, a post-hoc test after Kruskal-Wallis is adopted to find out the specific differences between the three achieving groups. As shown in Table 5, we can find that, *authority*, *betweenness*, *closeness*, *degree*, *eigenvector*, *indegree* and *pagerank* are all significantly different between high- and middle -achieving learners, and especially between high- and low -achieving learners. However, middle- and low-achieving learners do not significantly differ in these seven social characteristics. This can also be seen from Figure 3, in which there is no obvious difference in the distribution of middle- and low-achieving learners in the network. These results further suggested that, the position of learners in the network is correlated with their academic performance, as well as high-achieving group has a more active participation in discussions than middle- and low-achieving groups.

5. Conclusion

In this paper, we have utilized the social network analysis to investigate the differences of social characteristics between different achievement levels of learners within a course forum in a university online learning platform. To begin with, we demonstrated that most social characteristics have significantly positive correlations with academic achievements. In addition, we conducted a linear regression analyses and indicated that *eigenvector* centrality seemed to be a valid indicator to predict academic achievements, that is, the more positively the learners participate in the forum and communicate with the high-centrality peers, the better his/her academic achievement will be. Moreover, the distributions of network positions of various achieving learners show that the majority of high-achieving learners are located at the core positions in the network, while most middle-achieving and low-achieving learners are both located at the edge positions of the network. This reveals that a higher social centrality may be associated to a higher academic performance [19]. In the end, The Kruskal-Wallis H test and Pairwise Comparisons were used to examine the differences of social characteristics between various achieving learners, which reveals that high-achieving learners tend to be more active than middle-achieving and low-achieving learners in forum interactions (indicated by *authority*, *betweenness*, *closeness*, *degree*, *eigenvector*, *indegree* and *pagerank*). In future work, we will collect more large-scale data that covering more courses to get in-depth insights into the interaction learning patterns between online learners.

6. Acknowledgments

This work was supported by China Postdoctoral Science Foundation (Grant Nos. 30205010003), self-determined research funds of CCNU from the colleges' basic research and operation of MOE (Grant Nos. KJ02072017-2291), National Natural Science Foundation of China (Grant Nos. 11505071 and 61702207), and National Key Research and Development Program of China (Grant No. 2017YFB1401303).

References

[1] Xu L, Wang F and Yu B 2018 *Social Network Analysis of MOOC Learners Knowledge Building* (Singapore: Springer) pp 363-377
[2] Chen Y H and Chen P 2015 MOOC study group: Facilitation strategies, influential factors, and student perceived gains *Comput. Educ* **86** 5570
[3] Fox A 2013 From MOOCs to SPOCs. *Communications of the ACM* **56** 38-40
[4] Al-Rahmi W M, Alias N, Othman M S, Marin V I and Tur G 2018 A model of factors affecting learning performance through the use of social media in malaysian higher education *Comput. Educ* **121** 59-72
[5] Haythornthwaite C 1996 Social network analysis: an approach and technique for the study of information exchange *Libr. Infor. Sci. Res* **18** 323-342

[6] Wey T, Blumstein D T, Shen W and Jordan F 2008 Social network analysis of animal behaviour: a promising tool for the study of sociality *Anim. Behav* **75** 333-344
[7] Oleksandra P and Shane D 2016 Untangling MOOC learner networks *Int. Conf. on Learning Analytics and Knowledge (Edinburgh)* pp 208-212
[8] Wise A F, Cui Y and Jin W Q 2017 Honing in on social learning networks in MOOC forums: examining critical network definition decisions *Int. Conf. on Learning Analytics and Knowledge (Vancouver)* pp 383-392
[9] Liu C C, Chen Y C and Tai S J D 2017 A social network analysis on elementary student engagement in the networked creation community *Comput. Educ* **115** 114125
[10] Laningawijnen L, Ryan A M, Harakeh Z, Shin H and Vollebergh W A M 2018 The moderating role of popular peers achievement goals in 5th- and 6th-graders achievement-related friendships: a social network analysis *J. Educ. Psychol* **110** 289
[11] Williams E A, Zwolak J P, Dou R and Brewe E 2017 Engagement, integration, involvement: supporting academic performance and developing a classroom social network *preprint* arXiv:1706.04121
[12] Gitinabard N, Xue L, Lynch C F, Heckman S and Barnes T 2017 A social network analysis on blended courses *preprint* arXiv:1709.10215
[13] Lee J and Bonk C J 2016 Social network analysis of peer relationship and online interactions in a blended class using blogs *Internet. High. Educ* **28** 35-44
[14] Liu Z, Kang L Y, Domanska M, Liu S and Fang C L 2018 Social characteristics of Learners in a Course Forum and Their Relationship to Academic achievements *Int. Conf. on Computer Supported Education (Madeira)* pp 15-21
[15] Houston I, S L, Brady K, Narasimham G and Fisher D 2017 Pass the Idea Please: The Relationship between Network Position, Direct Engagement, and Course Performance in MOOCs *In Proc of the 4th ACM Conf on Learning @ Scale (Cambridge)* pp 295-298
[16] Joksimovi S, Manataki A, Gaevi D, Dawson S, Kovanovi V and De Kereki I F 2016 Translating network position into performance: importance of centrality in different network configurations *Int. Conf. on Learning Analytics and Knowledge (Edinburgh)* pp 314-323.
[17] Kelley T L 1939 The selection of upper and lower groups for the validation of test items. *J. Educ. Psychol* **30** 17-24
[18] Simmel G and Wolff K H 1950 *The Sociology of Georg Simmel* (Free Press)
[19] Bakerdoyle K 2013 *Social network analysis and education: theory, methods and applications* (Sage Publications)